SIMPLY
CHEMISTRY

CONSULTANT
Andy Extance worked for six-and-a-half years in early stage drug-discovery research before becoming a full-time science writer. His writing explores everything related to chemistry, from Earth's environment to space, from food to fusion, and from solar cells to how we smell.

CONTRIBUTORS
Victoria Atkinson is a freelance science writer. She writes mostly about chemistry – covering the latest research for publications such as *Chemistry World* and *Live Science* – and shares her enthusiasm for science with audiences of all ages through her work on popular-science books.

Kat Day holds a PhD in Chemistry from the University of Nottingham and taught A level chemistry and GCSE sciences for over 15 years. These days she works as a professional editor and writer of both non-fiction and fiction.

Rebecca Mileham studied physics and then science communication at Imperial College London before working for 10 years at the Science Museum, London. She now provides text for international exhibitions and writes contemporary science books on subjects including chemistry, physics, technology, and culture.

CONTENTS

7   **WHAT IS CHEMISTRY?**

ATOMS

10   **THE ESSENTIAL ATOM**
Atomic structure

11   **ELECTRON CLOUDS**
Atomic orbitals

12   **SHAPE-SHIFTING SUBSTANCES**
States of matter

14   **NATURE'S CHEAT SHEET**
The periodic table

16   **THE COSMIC FORGE**
Nucleosynthesis

17   **THE MAKEUP OF THE UNIVERSE**
Elemental abundance

18   **CHEMICAL PERSONALITIES**
Elemental properties

19   **ATOMIC SIBLINGS**
Isotopes

20   **THE LIMITS OF MATTER**
Hunting new elements

21   **SEEKING STABILITY**
Radioactive elements

22   **HOW MUCH IS THAT?**
Absolute amounts

24   **POWERS OF TEN**
Scientific notation

CHEMICALS

28   **INGREDIENTS OF A COMPOUND**
Chemical formulae

29   **THE CHEMISTRY CODE**
Balanced equations

30   **METALS ALONE AND TOGETHER**
Metals and alloys

31	**EXPLOSIVE ELEMENTS** *Alkali metals*		44	**LEFT- AND RIGHT-HANDED MOLECULES** *Stereochemistry*
32	**ELECTRON GRABBERS** *Halogens*		45	**REGULAR BEAUTIES** *Crystals and gems*
33	**HIGHLY VERSATILE** *Transition metals*		46	**WHEN OPPOSITES INTERACT** *Acids, bases, and salts*
34	**WIN OR LOSE** *Ions*		47	**MEASURING ACIDITY** *The pH scale*
35	**UNTOUCHABLE ELEMENTS** *The noble gases*		48	**CYCLIC STABILITY** *Aromatic molecules*
36	**ATOMS ASSEMBLE!** *Covalent and ionic bonding*		49	**GREEDY ATOMS** *Free radicals*
38	**MOLECULAR BLENDS IN ACTION** *Solutions and other mixtures*			

CHEMISTRY PROCESSES

40	**ELECTRONIC TUG OF WAR** *Electronegativity*			
41	**MOLECULAR MAGNETS** *Hydrogen bonds*		52	**DRIVING DISORDER** *Entropy*
42	**ATTRACTION EVERYWHERE** *Van der Waals forces*		53	**BOND ENERGIES** *Enthalpy*
43	**THE LEAGUE OF ELEMENTS** *Reactivity series*		54	**ENERGY IN AND OUT** *Endothermic and exothermic reactions*
			55	**ACCOUNTING FOR ATOMS** *Reactions and mass balance*
			56	**SWAPPING PARTNERS** *Chemical reactions*
			57	**FORWARDS AND BACKWARDS** *Reversible reactions*
			58	**WILL REACTIONS HAPPEN?** *Gibbs free energy*
			59	**CHEMISTRY'S INVISIBLE BARRIER** *Activation energy*
			60	**MAKING AN IMPACT** *Collision theory*

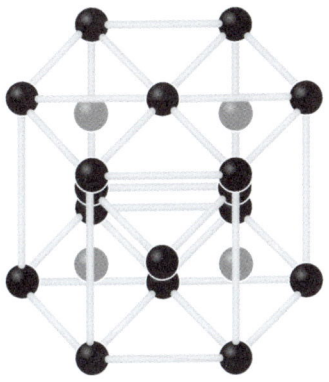

62 **FINDING BALANCE**
Equilibria and le Chatelier's principle

64 **CHEMICAL REACTION TYPES**
Addition, elimination, and substitution

65 **PUSHING TO COMPLETION**
Limiting reactants

66 **OVERWHELMING OXYGEN**
Oxidation and corrosion

67 **GETTING METAL**
Reduction from oxides

68 **FIRE, FURY, AND REACTIONS**
Combustion and explosions

69 **DRIVING REACTIONS WITH ELECTRICITY**
Electrolysis

70 **THE SILENT MATCHMAKERS**
Catalysts

71 **PURIFYING EARTH'S TREASURE**
Metal refining

72 **BUILDING SUPERMOLECULES**
Addition polymerization

73 **MAKING MOLECULAR GIANTS**
Condensation polymerization

74 **REACTIONS IN CONTINUOUS MOTION**
Flow chemistry

75 **HEAPS OF HYDROGEN**
Hydrogenation

76 **SEPARATING SUBSTANCES WITH HEAT**
Fractional distillation

77 **COLOURFUL CLOCKS**
Chemical oscillators

ANALYTICAL CHEMISTRY

80 **PULLING MIXTURES APART**
Chromatography

81 **BURNING BRIGHTLY**
Flame tests

82 **DETECTIVE WORK**
Chemical analysis tests

84 **FINGERPRINTING ELEMENTS**
Emission spectroscopy

85 **MOLECULAR MOTION**
Vibrational spectroscopy

86 **WEIGHING THE INVISIBLE**
Mass spectrometry

87 **NUCLEI IN MAGNETIC FIELDS**
Nuclear magnetic resonance

88 **MAPPING SURFACES**
Atomic force microscopy

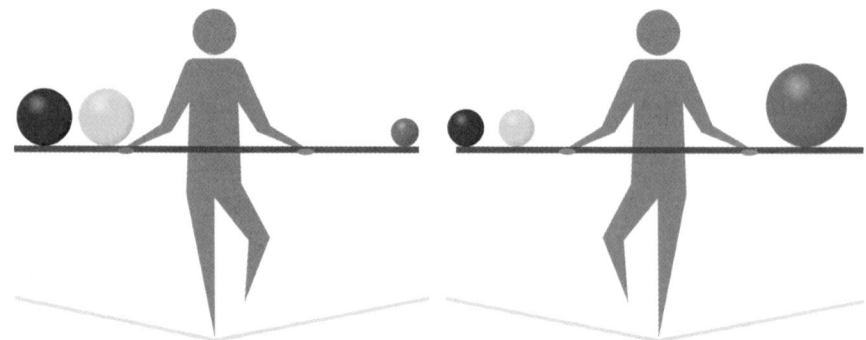

89	**FREEZING VISION** Cryo-electron microscopy	105	**NATURE'S PERFECT CATALYSTS** Enzymes
90	**EXAMINING CRYSTALLINE SUBSTANCES** X-ray crystallography	106	**LIFE'S GENETIC SPIRAL** DNA
91	**GAS SLEUTHING** Gas chromatography	107	**CELLS' ENERGY CURRENCY** ATP

BIOORGANIC CHEMISTRY

94	**THE SMART CHEMISTRY REVOLUTION** AI in chemistry
95	**LOTS OF LINKAGES** Single, double, and triple bonds
96	**VARYING FORMS** Allotropes
98	**THE CARBON ZOO** Alkanes, alkenes, and alkynes
100	**BREAKING BONDS** Cracking alkanes to alkenes
101	**REACTIVE HYDROCARBONS** Alkene reactions
102	**BUILDING BLOCKS OF LIFE** Amino acids and proteins
103	**NATURE'S SWEET FUEL** Carbohydrates
104	**BIOLOGY'S BARRIERS** Lipids

108	**TRANSPORTING OXYGEN** Haemoglobin
109	**CHEMICAL SIGNALLERS** Hormones
110	**MOLECULES OF CONSCIOUSNESS** Neurotransmitters
111	**CLEANING UP** Surfactants
112	**LIVING FACTORIES** Biosynthetic pathways
113	**TINY GEARS IN ACTION** Molecular machines

NATURAL CHEMISTRY

116	**THE MOLECULE OF LIFE** The properties of water
117	**TURNING LIGHT INTO LEAVES** Photosynthesis
118	**FROM HELL TO EARTH** The early atmosphere

119 **BREATHING CHEMISTRY**
Modern atmospheric composition

120 **TRAPPING CLIMATE CULPRITS**
Greenhouse gases and carbon capture

122 **HARMING EARTH'S SHIELD**
The ozone layer and CFCs

123 **MOLECULES BEYOND EARTH**
Chemicals in space

INDUSTRIAL CHEMISTRY

126 **MAKING AMMONIA**
The Haber process

127 **POWERING PLANTS**
Making fertilizers

128 **MOULDABLE MARVELS**
Plastics

130 **CALLING TIME ON FOSSIL FUELS**
Fossil fuels

131 **CLEANER POWER**
Alternative fuels

132 **SPLITTING WATER**
Green hydrogen

133 **MOLECULAR RENEWAL**
Recycling plastic

134 **CHEMISTRY VS. NATURE'S PESTS**
Herbicides and pesticides

135 **INDESTRUCTIBLE POLLUTANTS**
Forever chemicals

136 **CLEAN ENOUGH TO DRINK**
Water sterilization

137 **TOXIC BUT ESSENTIAL**
Heavy metals

138 **ATOMIC LEGACY**
Nuclear waste

139 **TINY PILLS, HUGE IMPACT**
Small molecule drugs

140 **DEFENDING WITH CHEMISTRY**
Vaccines

141 **PRECISION IMMUNE WARFARE**
Antibody drugs

142 **THE SUPERBUG BATTLE**
Antibiotic resistance

143 **READING LIFE'S CODE**
DNA sequencing

144 **SHOW WITH THE FLOW**
Lateral flow tests

145 **FIXING GENES WITH CHEMISTRY**
DNA therapy

146 **REWRITING LIFE'S CODE**
CRISPR gene editing

148 **SUPER SWITCHES**
Semiconductors

149 **DISPLAY STRUCTURES**
Liquid crystals

150 **BRIGHTER, GREENER SCREENS**
Organic light-emitting diodes

151 **TINY TECH, HUGE IMPACT**
Quantum dots

152 **CHEMICAL SCRUBBING**
Removing exhaust gases

153 **TINY HOLES, BIG USES**
Porous materials

154 **POWERING THE FUTURE**
Lithium-ion batteries

155 **PORTABLE ENERGY**
Fuel cells

156 **INDEX**

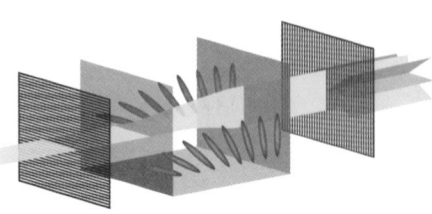

WHAT IS CHEMISTRY?

You are! People are built from elements: mostly oxygen, carbon, hydrogen, and nitrogen. And you have a lot in common with the paper in a book – it is a network of carbon, hydrogen, and oxygen atoms, together called cellulose. Everything we see is chemistry, as are some things we do not see. The air is mostly nitrogen molecules, some oxygen molecules to breathe, and more of the greenhouse gas carbon dioxide than scientists think is safe.

Chemistry is also about how arranging and connecting elemental building blocks differently can create such varied entities as a person, a book, and air. It is about how and why some atoms of elements connect with each other and with other elements. It is also about why some elements are reluctant to interact at all. Chemistry is what separates inactive substances, like dust and rocks, from living organisms, which are animated by complex interactions.

Such knowledge has improved life for many. But it has also caused pollution that counteracts those benefits in numerous ways. Chemistry could remedy the problems caused by pollution – but it must be used wisely.

ATOMS

Every substance, whether solid, liquid, or gas, is made of atoms. These particles are so minuscule that a sheet of paper is up to one million atoms thick. Nearly all the mass of an atom is concentrated in its nucleus, which is a part of the atom one ten-thousandth of its overall size. An atom's identity comes from the number of protons in its nucleus. The periodic table arranges the elements in order of increasing proton number, starting at 1 with hydrogen. Currently, the table finishes at oganesson, which has 118 protons, but scientists hope to produce even heavier elements in particle accelerators.

THE ESSENTIAL ATOM

From life-giving oxygen to deadly arsenic, atoms of every substance are made of just three components: protons, neutrons, and electrons. At the centre of each atom sits a tiny nucleus, made of positively charged protons and neutrons with no charge. A cloud of negatively charged electrons surrounds the nucleus, giving the atom an overall neutral charge. Since protons and neutrons are over 1,800 times heavier than electrons, the nucleus accounts for the majority of an atom's mass.

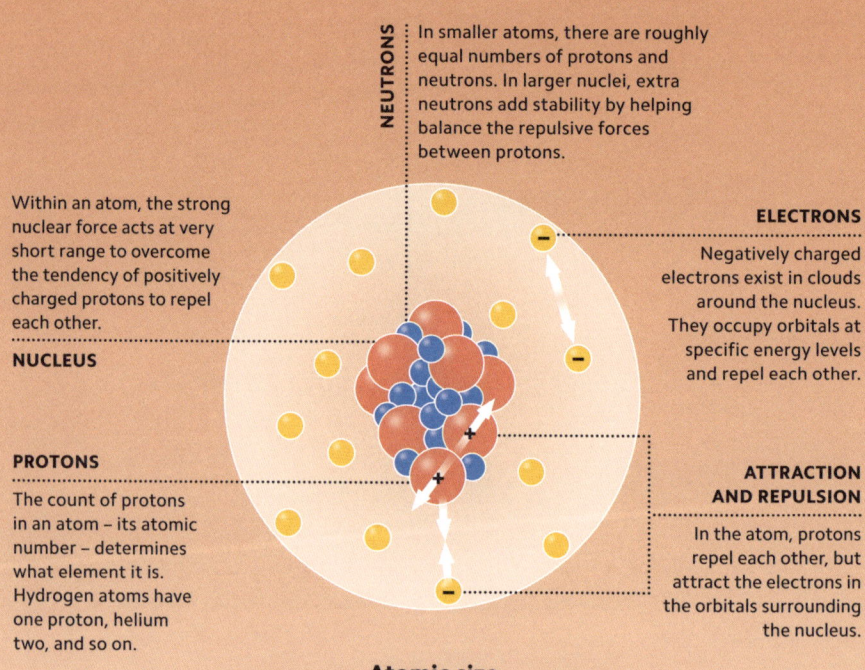

NEUTRONS
In smaller atoms, there are roughly equal numbers of protons and neutrons. In larger nuclei, extra neutrons add stability by helping balance the repulsive forces between protons.

Within an atom, the strong nuclear force acts at very short range to overcome the tendency of positively charged protons to repel each other.

NUCLEUS

ELECTRONS
Negatively charged electrons exist in clouds around the nucleus. They occupy orbitals at specific energy levels and repel each other.

PROTONS
The count of protons in an atom – its atomic number – determines what element it is. Hydrogen atoms have one proton, helium two, and so on.

ATTRACTION AND REPULSION
In the atom, protons repel each other, but attract the electrons in the orbitals surrounding the nucleus.

Atomic size
An atom's radius is about a tenth of a billionth of a metre. If an atom was scaled up to the size of a football stadium, its nucleus would be about the size of a grape.

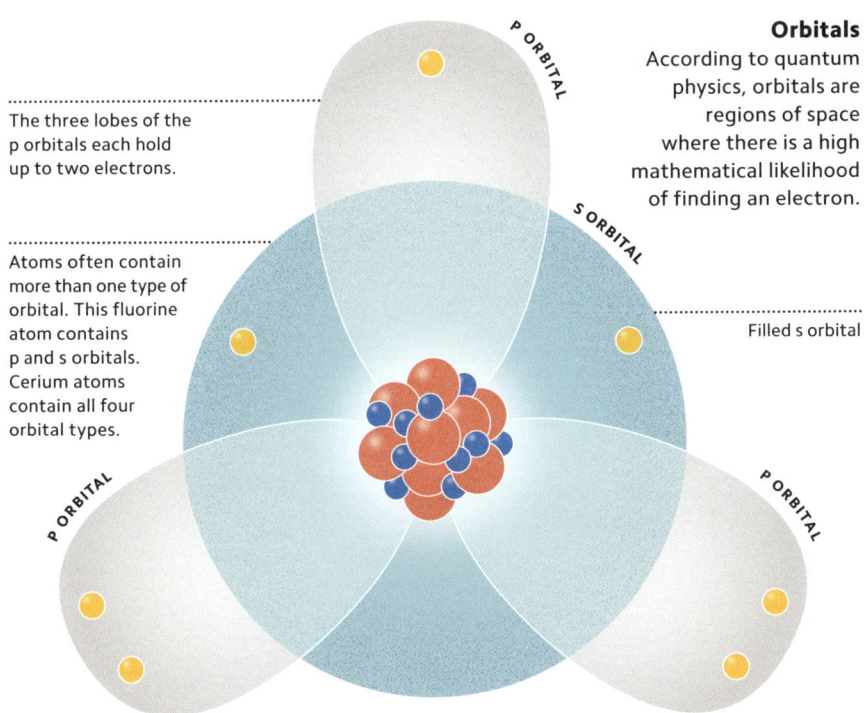

Orbitals
According to quantum physics, orbitals are regions of space where there is a high mathematical likelihood of finding an electron.

The three lobes of the p orbitals each hold up to two electrons.

Atoms often contain more than one type of orbital. This fluorine atom contains p and s orbitals. Cerium atoms contain all four orbital types.

Filled s orbital

ELECTRON CLOUDS

Scientists once thought that electrons circled an atom's nucleus like tiny planets. Today, quantum theory shows us that electrons are actually found in balloon-shaped areas of space called orbitals. Spherical s orbitals surround the nucleus, at the lowest energy level, holding up to two electrons. At higher energy levels, p orbitals, which are sets of three figure-of-eight shapes, or lobes, contain in total up to six electrons. Larger, multi-lobed d and f orbitals can hold up to 10 and 14 electrons respectively. The atom's outermost electrons, called valence electrons, define its chemical behaviour.

ATOMIC ORBITALS | 11

INCREASING ENERGY

LIQUIDS
A liquid forms when a solid is heated beyond its melting point. Intermolecular forces are still present, but the particles can move relative to each other, allowing the liquid to flow.

ENERGY
To change a substance from solid to liquid to gas – or back – requires an input or removal of energy.

MELTING *FREEZING* *CONDENSATION* *EVAPORATION*

Changing state
By changing the pressure or temperature, the state of a substance can be changed. The changes can be reversed and the matter is not chemically altered.

Some gases can bypass the liquid state when becoming a solid.

DEPOSITION

SUBLIMATION

Atoms or molecules

Some solids change to gas directly, bypassing the liquid state.

SOLIDS
A solid's particles are closely packed together and bound by intermolecular forces or chemical bonds. Solids maintain a fixed shape and volume.

GASES
In a gas, the particles are too far apart for intermolecular forces to act, so the particles will move about. They expand to fill a container.

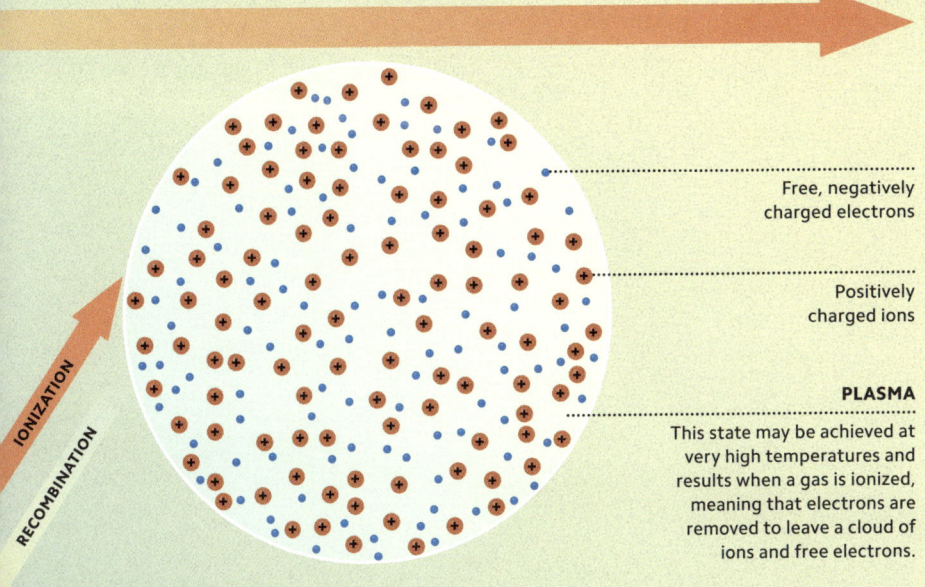

Free, negatively charged electrons

Positively charged ions

PLASMA
This state may be achieved at very high temperatures and results when a gas is ionized, meaning that electrons are removed to leave a cloud of ions and free electrons.

SHAPE-SHIFTING SUBSTANCES

Solid, liquid, gas, or plasma – a substance exists in different states depending on temperature and pressure. The closely-packed atoms in a solid material are held together by intermolecular forces (forces between molecules). When heated, the particles in the solid vibrate and may overcome the forces between them, melting into a liquid. If more energy is added, the liquid boils and disperses as a gas. At very high temperatures, a gas may turn into a plasma, seen in nature as flames and lightning, and in auroras in Earth's sky and stars. In a pure material, the temperature at which changes of state occur is a key way to identify the substance.

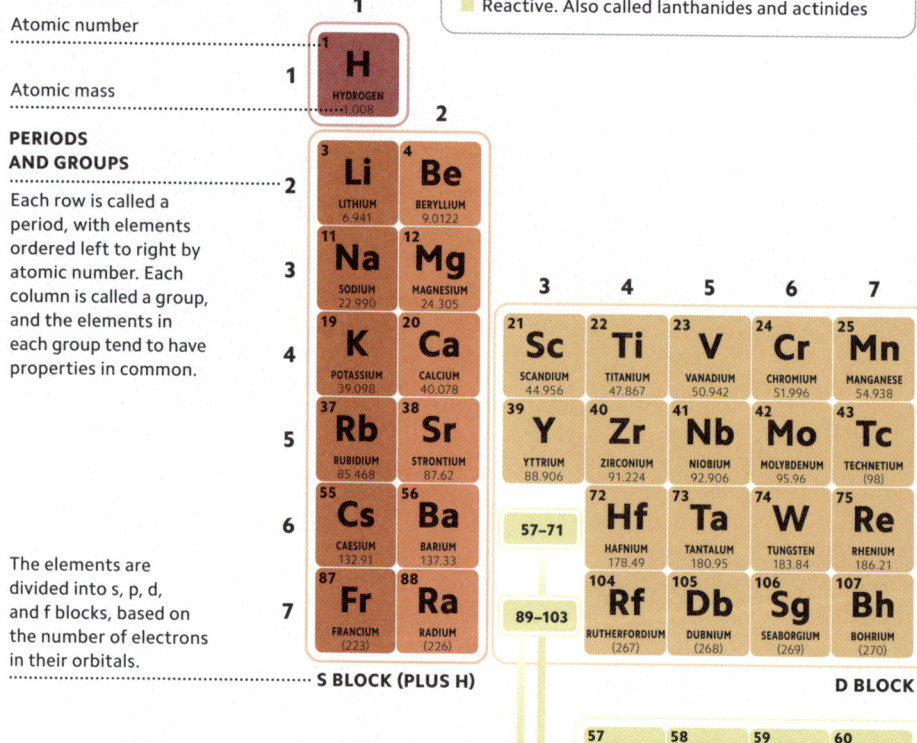

KEY
REACTIVE METALS
■ **Alkali metals** – these elements have one valence electron, making them highly reactive and ready to form compounds
■ **Alkaline earth metals** – moderately reactive metals

TRANSITION ELEMENTS
■ **Transition metals** – a varied group, many of which have valuable properties

MAINLY NONMETALS
■ **Metalloids** – possess properties between those of metals and nonmetals
■ **Other metals** – mostly relatively soft metals with low melting points
■ **Carbon** and other nonmetals
■ **Halogens** – very reactive nonmetals; seven valence electrons each
■ **Noble gases** – have a full outer electron shell; are almost entirely inert

RARE EARTH METALS
■ Reactive. Also called lanthanides and actinides

Atomic number
Atomic mass

PERIODS AND GROUPS

Each row is called a period, with elements ordered left to right by atomic number. Each column is called a group, and the elements in each group tend to have properties in common.

The elements are divided into s, p, d, and f blocks, based on the number of electrons in their orbitals.

In these elements, some of which are radioactive and synthetic, the 4f and 5f orbitals gradually fill and can hold up to 14 electrons. They are shown separately from the main table for visual compactness.

LANTHANIDES AND ACTINIDES

14 | THE PERIODIC TABLE

NATURE'S CHEAT SHEET

Each chemical element is defined by the number of protons in its atomic nucleus. The periodic table, devised by Russian chemist Dmitri Mendeleev in the 1860s, shows all known pure substances in horizontal rows (periods) of increasing proton number. Proton numbers increase from left to right across the periods, along with the number of electrons, which gradually fill the orbitals (see p.11). The elements in the columns (groups) have similar chemical properties as they have the same number of valence electrons (see p.11) available for bonding and reactions.

The right-hand side of the table features nonmetals, including halogens, noble gases, and key elements for life, such as carbon, nitrogen, oxygen, and sulfur. Most are gases at room temperature.

NONMETALS

8	9	10	11	12
26 **Fe** IRON 55.845	27 **Co** COBALT 58.933	28 **Ni** NICKEL 58.693	29 **Cu** COPPER 63.546	30 **Zn** ZINC 65.38
44 **Ru** RUTHENIUM 101.07	45 **Rh** RHODIUM 102.91	46 **Pd** PALLADIUM 106.42	47 **Ag** SILVER 107.87	48 **Cd** CADMIUM 112.41
76 **Os** OSMIUM 190.23	77 **Ir** IRIDIUM 192.22	78 **Pt** PLATINUM 195.08	79 **Au** GOLD 196.97	80 **Hg** MERCURY 200.59
108 **Hs** HASSIUM (277)	109 **Mt** MEITNERIUM (278)	110 **Ds** DARMSTADTIUM (281)	111 **Rg** ROENTGENIUM (282)	112 **Cn** COPERNICIUM (285)

13	14	15	16	17	18
					2 **He** HELIUM 4.003
5 **B** BORON 10.81	6 **C** CARBON 12.011	7 **N** NITROGEN 14.007	8 **O** OXYGEN 15.999	9 **F** FLUORINE 18.998	10 **Ne** NEON 20.180
13 **Al** ALUMINIUM 26.982	14 **Si** SILICON 28.085	15 **P** PHOSPHORUS 30.974	16 **S** SULFUR 32.06	17 **Cl** CHLORINE 35.45	18 **Ar** ARGON 39.948
31 **Ga** GALLIUM 69.723	32 **Ge** GERMANIUM 72.63	33 **As** ARSENIC 74.922	34 **Se** SELENIUM 78.971	35 **Br** BROMINE 79.904	36 **Kr** KRYPTON 83.798
49 **In** INDIUM 114.82	50 **Sn** TIN 118.71	51 **Sb** ANTIMONY 121.76	52 **Te** TELLURIUM 127.60	53 **I** IODINE 126.90	54 **Xe** XENON 131.29
81 **Tl** THALLIUM 204.38	82 **Pb** LEAD 207.20	83 **Bi** BISMUTH 208.98	84 **Po** POLONIUM (209)	85 **At** ASTATINE (210)	86 **Rn** RADON (222)
113 **Nh** NIHONIUM (286)	114 **Fl** FLEROVIUM (289)	115 **Mc** MOSCOVIUM (289)	116 **Lv** LIVERMORIUM (293)	117 **Ts** TENNESSINE (294)	118 **Og** OGANESSON (294)

P BLOCK (EXCLUDING He)

61 **Pm** PROMETHIUM (145)	62 **Sm** SAMARIUM 150.36	63 **Eu** EUROPIUM 151.96	64 **Gd** GADOLINIUM 157.25	65 **Tb** TERBIUM 158.93	66 **Dy** DYSPROSIUM 162.50	67 **Ho** HOLMIUM 164.93	68 **Er** ERBIUM 167.26	69 **Tm** THULIUM 168.93	70 **Yb** YTTERBIUM 173.05	71 **Lu** LUTETIUM 174.97
93 **Np** NEPTUNIUM (237)	94 **Pu** PLUTONIUM (244)	95 **Am** AMERICIUM (243)	96 **Cm** CURIUM (247)	97 **Bk** BERKELIUM (247)	98 **Cf** CALIFORNIUM (251)	99 **Es** EINSTEINIUM (252)	100 **Fm** FERMIUM (257)	101 **Md** MENDELEVIUM (258)	102 **No** NOBELIUM (259)	103 **Lr** LAWRENCIUM (262)

F BLOCK

Red supergiant stars create carbon, oxygen, magnesium, and other elements by fusing nuclei together.

Concentric shells develop within the supergiant star, each producing successively heavier nuclei.

When the nuclei of one element have fused, new layers form that fuse together the elements the stars had previously made.

Supergiant star
After exhausting the hydrogen fuel in their core, very large stars can swell into red supergiants, fusing lighter nuclei to produce elements as heavy as iron.

THE COSMIC FORGE

Nucleosynthesis is the process by which stars build elements, using hydrogen as their raw material. The stars are celestial element factories, fusing hydrogen into helium – a process in which four light atomic nuclei combine, through heat and pressure, to form a single, heavier nucleus. When the hydrogen is exhausted, the stars fuse helium into heavier elements, forming shells containing successively larger nuclei. At the end of a star's life, it explodes, blasting its wealth of nuclei out into space.

THE MAKEUP OF THE UNIVERSE

Over 98 per cent of the matter in the Universe is hydrogen and helium, as it was shortly after the Big Bang. The remaining two per cent consists of heavier elements blown outwards when old stars explode. Heavier elements may clump together to form planets. Earth's most abundant element is iron, which lies mostly in its core. Earth's less abundant elements, such as phosphorus, oxygen, nitrogen, carbon, hydrogen, and sulfur, support life. Other elements exist in varying quantities.

Hydrogen (73.9%)

Helium (24%)

CARBON (0.5%)
Elements with even atomic numbers are more abundant than those with odd atomic numbers, as it takes less energy for them to form.

Oxygen (1%)

ALL OTHERS (0.6%)
Elements such as lithium and boron are produced by fusion but destroyed by other stellar reactions – making them rare.

Abundance on Earth
Elements are not distributed evenly across the Universe. On Earth, hydrogen, helium, neon, and nitrogen are less abundant because they vaporize easily, and escape into space.

- Iron (32.1%)
- Oxygen (30.1%)
- Silicon (15.1%)
- Magnesium (13.9%)
- Sulfur (2.9%)
- Nickel (1.8%)
- Calcium (1.5%)
- Aluminium (1.4%)
- Trace elements (1.2%)

ELEMENTAL ABUNDANCE

CHEMICAL PERSONALITIES

Elements vary widely in their character and behaviour, largely due to their outermost, valence, electrons. These properties are reflected in the element's location on the periodic table (see pp.14–15). In metals, electrons become loose, or delocalize, allowing metallic elements to conduct electricity and to deform. Nonmetals hold on to their valence electrons tightly, and are brittle and non-conductive. Semimetals are somewhere in between. Metals are shiny because the free-moving sea of electrons reflects a broad spectrum of visible light. By comparison, nonmetals are dull.

3 OR LESS VALENCE ELECTRONS
Metals have up to three valence electrons, and give electrons away to form ionic bonds (see p.37). Electrons are shared collectively.

LARGE ATOMIC RADIUS
Electrons are loosely held by, and therefore further away from, the nuclei of metal atoms, resulting in a larger atomic radius.

LOSING ELECTRONS
Down a group, the outer electrons are further from the nucleus and weakly held. Less energy is needed to detach an electron, ionizing the atom.

METALS
NONMETALS

5 TO 7 VALENCE ELECTRONS
Nonmetals have five to seven valence electrons, which remain bound to specific atoms. This makes them poor conductors.

SMALL ATOMIC RADIUS
Nonmetals have a smaller atomic radius due to the strong pull the nucleus has on the electrons, holding on to them tightly.

ATTRACTING ELECTRONS
Nuclei size increase across a period, attracting electrons more strongly. Nonmetals can gain further electrons and become an anion.

18 | ELEMENTAL PROPERTIES

ATOMIC SIBLINGS

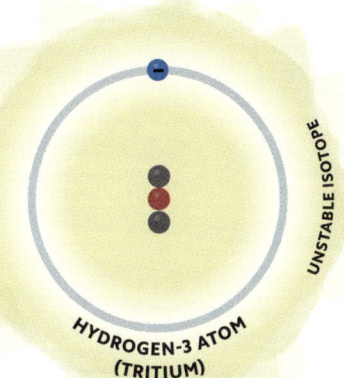

The word isotope comes from the Greek words *isos*, meaning "same", and *topos*, meaning "place". This is because it refers to atoms that have the same number of protons – and hence the same elementary identity and position on the periodic table – but a different number of neutrons in their nucleus. Most elements have more than one naturally occurring isotope. The variants are denoted by their mass number, which is the number of protons and neutrons added together.

Hydrogen and deuterium formed just after the Big Bang, about 13.8 billion years ago.

Tritium forms in trace amounts through cosmic ray interaction with gases in the atmosphere. It decays again quickly because its nucleus is unstable (see p.21).

Three hydrogens
Hydrogen has three naturally occurring isotopes, with zero, one, and two neutrons. This gives them a mass number of 1, 2, and 3 respectively.

ISOTOPES | 19

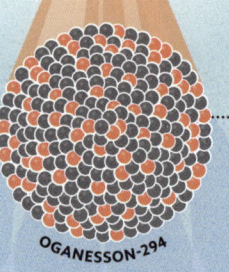

COLLISION
Bombarding californium atoms (98 protons) with calcium ions (20 protons) can fuse them into oganesson (118 protons).

On the very rare occasion, californium and calcium nuclei fuse, three neutrons are released during the reaction.
RELEASE

Rare element
Only five or six atoms of oganesson have been created in particle accelerators since its discovery in 2002.

CREATION
From this nuclear reaction, the element oganesson forms, lasting less than a millisecond before it decays radioactively.

THE LIMITS OF MATTER

The elements up to plutonium (94 on the periodic table) occur naturally. Beyond this, elements are made synthetically in a nuclear reactor or particle accelerator. Most exist only fleetingly, in minuscule amounts. The heaviest known element so far is oganesson (118), produced by bombarding a target of californium-249 with ions of calcium-48 to create a nucleus with the atomic weight 294. Scientists hope to discover even heavier elements, that, unlike the short-lived oganesson, may even prove to be stable.

SEEKING STABILITY

Elements whose nuclei are very large or have an unstable balance of protons and neutrons are said to be radioactive. These nuclei spontaneously emit a form of energy called gamma radiation or release particles to achieve stability in a process called radioactive decay. Many of the most familiar elements have at least one radioactive isotope, such as carbon-14, for example.

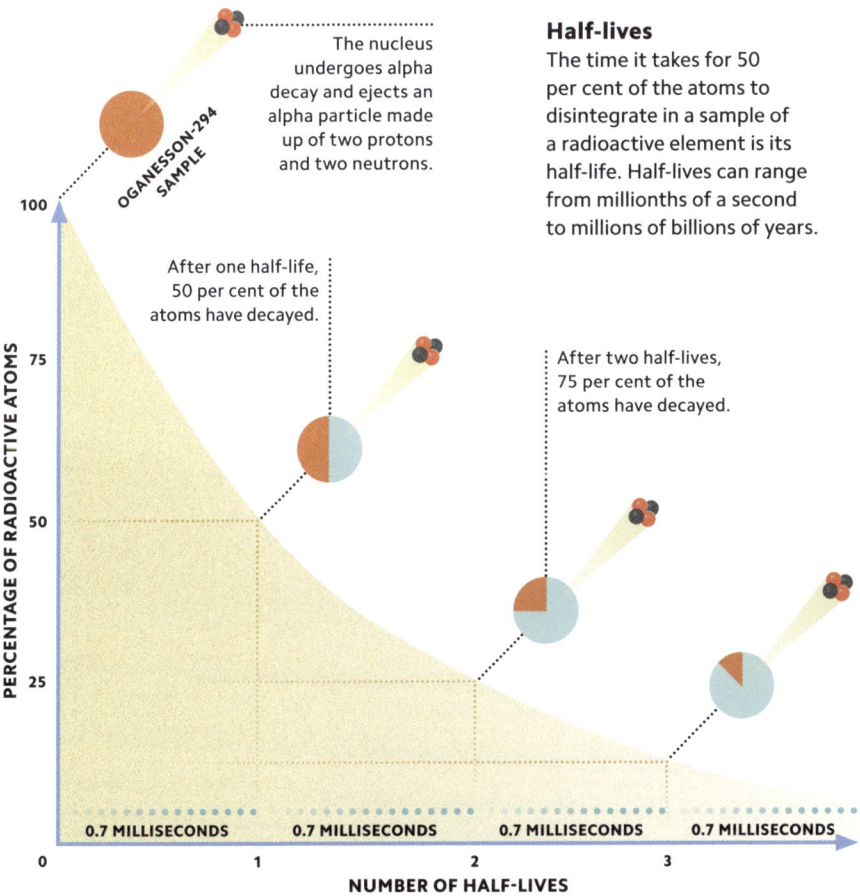

Atoms are so small that it is convenient to refer to them in bulk. The unit of measurement used is called a mole, and one mole of any substance is equal to 6.022×10^{23} particles, be they atoms, molecules, or isotopes. This huge figure is known as Avogadro's number and it is how many atoms there are in 12 g (0.4 oz) of carbon-12, the benchmark by which all other substances are measured. Moles are used rather than grams to quantify substances in chemistry because different substances have different masses.

HOW MUCH IS THAT?

MASS GRAMS (g)

4 MOL OF CO_2 CO_2

The number of moles in a substance multiplied by its molar mass give the mass of the substance.

The heavier the atoms are in a substance, the more one mole of it will weigh. There are the same number of atoms in 1 gram of hydrogen as in 197 grams of gold.

MOLES MOL

1 MOLE EQUALS:

6.022×10^{23} ATOMS

AND IS THE EQUIVALENT TO:

	H	He	C	Fe	Au
	HYDROGEN	HELIUM	CARBON	IRON	GOLD
Atomic mass	1.008	4.003	12.011	55.845	196.97
	1 g	**4 g**	**12 g**	**55.9 g**	**197 g**

22 | ABSOLUTE AMOUNTS

> A mole of grapefruits would be the size of Earth, but much lighter.

CARBON DIOXIDE

CO_2 CO_2

Number of moles

4 **X** 44 **=**

176 g (6 oz)   Molar mass

Molar mass is the weight in grams of one mole of a substance. This substance could be a single element, or all the elements in a chemical formula when their atomic masses are added up.

Formula triangle
If two of three variables (mass, moles, or molar mass) in a formula are known, this triangle helps work out the third. For example, the number of moles present is found by dividing the mass by the molar mass.

MOLAR MASS
GRAMS PER MOL (g/mol)

CARBON
C

C
CARBON
12

↓

12

12 g/mol

CARBON DIOXIDE
CO_2

C   O   O
CARBON   OXYGEN   OXYGEN
12   16   16

↓   ↓   ↓

12 **+** 16 **+** 16 **=**

44 g/mol

ABSOLUTE AMOUNTS

Tiny to titan

Scientific notation can express the mass of things as small as an electron and as large as the Universe and all it encompasses.

10^{14}
The total live biomass on Earth, mostly plants, is about 5.5×10^{14} kg.

10^{-31}
The mass of an electron is 9.1×10^{-31} kg.

10^{8}
The asteroid that killed the dinosaurs brought 5×10^{8} kg of iridium to Earth, according to estimates.

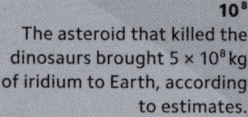

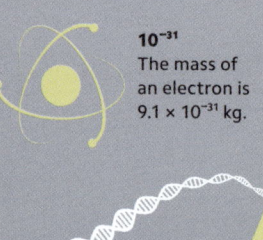

10^{-2}
There is around 50 g of DNA in a human body, which is 5×10^{-2} kg.

10^{0}
An average human body contains 1 kg (1×10^{0}) of calcium in the teeth and bones. Nitrogen, found in proteins, DNA, and RNA, is double this (2×10^{0} kg).

10^{10}

10^{-27}
The mass of an atom of hydrogen, the lightest element, is 1.673×10^{-27} kg.

10^{5}

10^{1}

10^{9} SECONDS
1.48×10^{9} seconds is 47 years – how long the Voyager 1 space probe has been travelling. It sent back data about the chemical atmospheres of planets.

10^{0}

10^{-10}

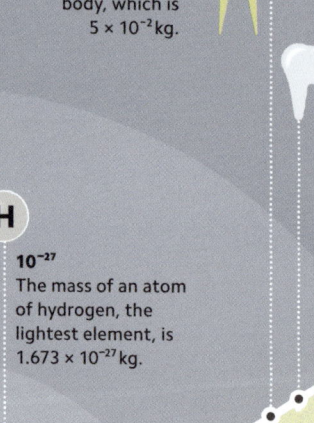

10^{-20}

10^{-32} SECONDS
The Universe underwent rapid expansion about 10^{-32} seconds after the Big Bang.

10^{-30}

SCIENTIFIC NOTATION

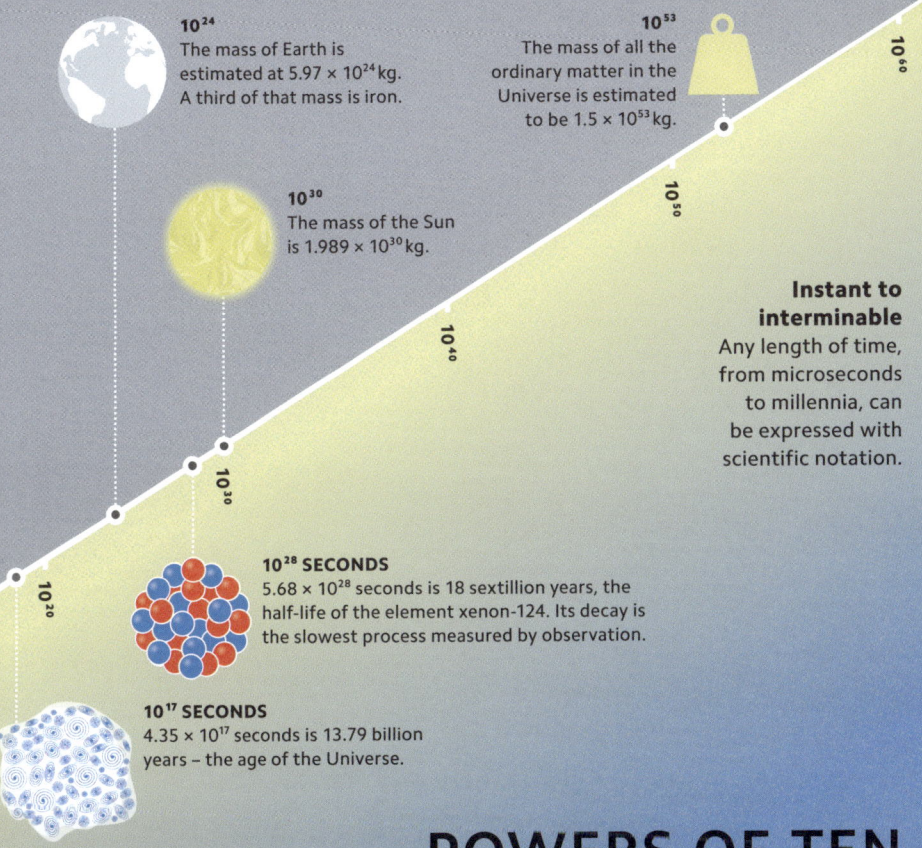

10^{24}
The mass of Earth is estimated at 5.97×10^{24} kg. A third of that mass is iron.

10^{53}
The mass of all the ordinary matter in the Universe is estimated to be 1.5×10^{53} kg.

10^{30}
The mass of the Sun is 1.989×10^{30} kg.

Instant to interminable
Any length of time, from microseconds to millennia, can be expressed with scientific notation.

10^{28} SECONDS
5.68×10^{28} seconds is 18 sextillion years, the half-life of the element xenon-124. Its decay is the slowest process measured by observation.

10^{17} SECONDS
4.35×10^{17} seconds is 13.79 billion years – the age of the Universe.

POWERS OF TEN

From the minuscule to the massive, the range over which chemistry operates can make lengths and masses cumbersome to write out. Scientists therefore often express numbers in scientific notation. This means turning a quantity into a number between 1 and 10, multiplied by ten raised to a whole-number power. For example, 500 is 5×10^2 ($5 \times 10 \times 10$), while 0.005 is 5×10^{-3} ($5 \div (10 \times 10 \times 10)$). When using this form of notation, the numbers can seem deceptively small and get very big very quickly. To represent time, the modest-looking 1×10^6 seconds represents 1 million seconds; 1×10^9 seconds is 1 billion seconds.

SCIENTIFIC NOTATION

CHEMI

CALS

The chemical behaviour of an atom is governed by the electrons in its outermost shell. Known as valence electrons, they are the furthest from the atom's nucleus and therefore the least tightly held. An atom's reactivity is determined by the number of its valence electrons, as chemists realized early in the 20th century. In the search to achieve a stable energy state, atoms gain, lose, or share their valence electrons. This opens them up to the possibility of bonding and forming compounds. This fundamental principle leads to the creation of all the diverse matter around us.

INGREDIENTS OF A COMPOUND

Different elements can react with one another to chemically bond together, making new structures called compounds, with new properties. Chemical formulae use letters and numbers to represent the elements in a compound, while diagrams show a compound's structure. The composition and structure of water can be shown in the following ways.

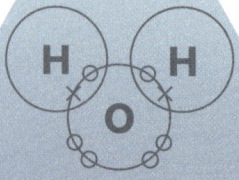

Dot and cross diagram
The outer shells of the atoms are represented by circles, with dots and crosses showing the electrons from each atom.

Chemical formula
H_2O
H is the symbol for hydrogen. O represents the oxygen atom.

A chemical formula uses atomic symbols to show the composition of each molecule of a substance, including the number of atoms of each element.

Visual formula
Two hydrogen atoms
One oxygen atom
The atoms of each element in the molecule and their bonds are visually represented by a simple outline.

Structural formula
H—O—H
The dashes represents the single bond between the atoms.

The atomic symbols are joined by lines representing the single, double, or triple bonds (see p.95) between the elements.

28 | CHEMICAL FORMULAE

THE CHEMISTRY CODE

Elements and their compounds can react together to form new products. Chemical equations represent reactions using symbols, showing the formulae of all the substances involved. The ingredients of the reaction are known as the reactants, and the resulting compounds are called the products. The arrow indicates the change from reactants to products in the chemical equation.

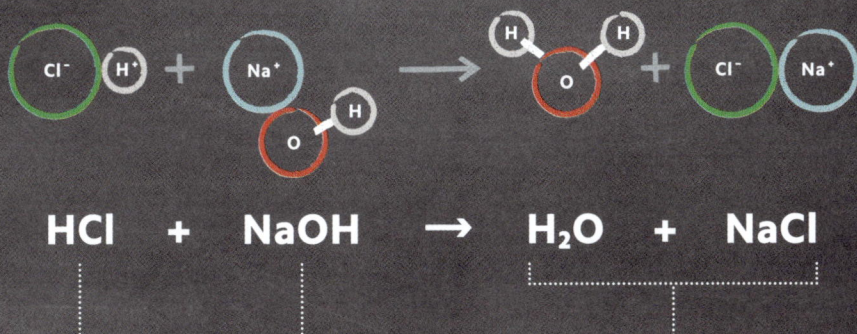

$$HCl + NaOH \rightarrow H_2O + NaCl$$

In hydrochloric acid (HCl) there is one hydrogen ion and one chlorine ion.

Sodium hydroxide (NaOH) molecules contain one sodium ion along with one oxygen atom and one hydrogen atom covalently bonded together in a hydroxide ion.

When the two compounds react, they form sodium chloride (NaCl), with one ion each of sodium and chlorine, and water (H_2O), which contains two hydrogen atoms and one oxygen atom.

Chemical equations

When elements or compounds undergo a chemical reaction, the atoms rearrange to form a new product, but the total number of atoms remains unchanged. A balanced chemical equation has the same number of atoms on both sides.

BALANCED EQUATIONS | 29

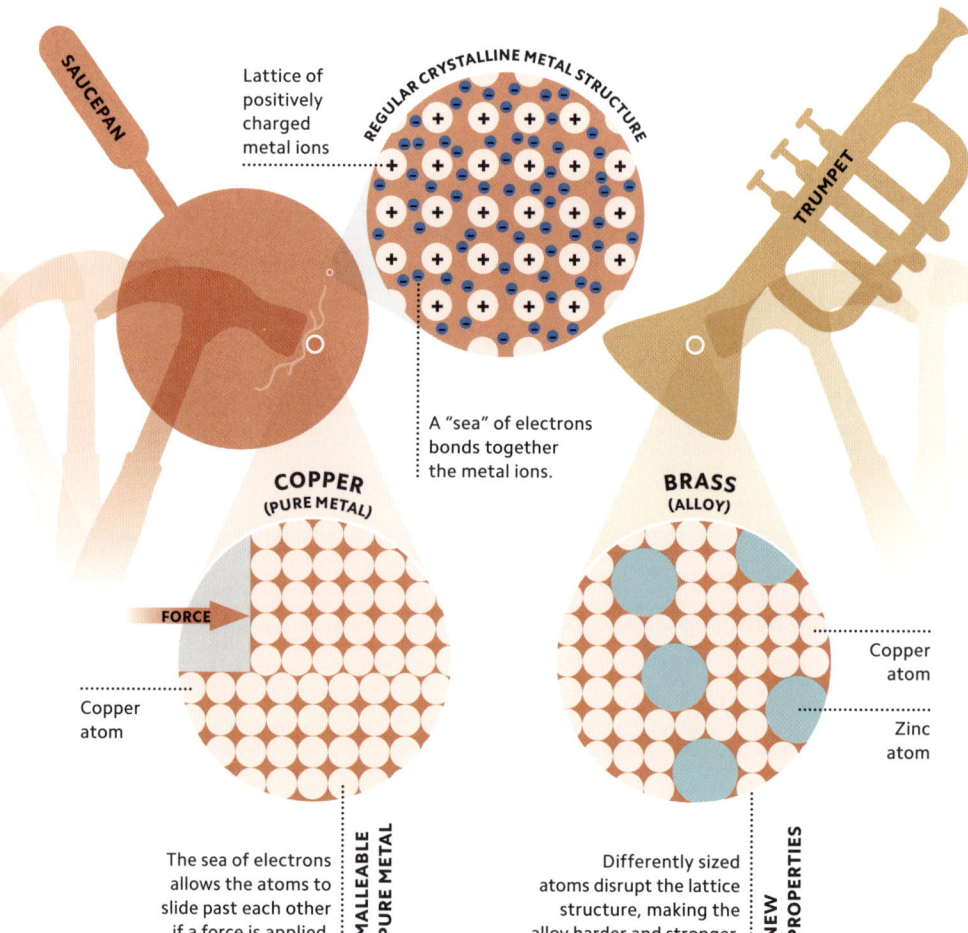

METALS ALONE AND TOGETHER

Most elements are metals – materials that are usually shiny, strong, and good conductors of heat and electricity. In metals, the atoms are packed closely together. Their valence electrons form a shared "sea" that surrounds the metal ions and holds them together. Adding a second element to the metal can create a mixture called an alloy. The alloying atoms substitute for some of the original atoms, or sit in between them.

EXPLOSIVE ELEMENTS

The alkali metals occupy the first column of the periodic table (see pp.14–15). They all have one electron in the outermost (valence) shell. To create a complete outer shell that is more stable, they lose this valence electron, which makes them highly reactive. They are never found uncombined in nature, and in the lab, they are stored in oil or in a vacuum to avoid any reactions.

ELECTRON SHELL
The outermost shell of alkali metals has just one electron, called a valence electron.

NUCLEUS
As the number of electrons around the nucleus increases, they counteract the attractive force of the nucleus, and it becomes easier to lose the outermost electron.

High reactivity
Alkali metals are silvery and soft in their pure form. They react vigorously with water to form hydrogen gas and with oxygen in the air to form oxides.

ALKALI METALS | 31

The halogen group
Iodine is the fourth element in the halogens group, so it is less reactive than fluorine, chlorine, and bromine but more reactive than astatine. The reactivity of the last element in the group, tennessine, is unknown.

The outermost shell of all halogen atoms has seven electrons.
INCOMPLETE SHELL

ADDING AN ELECTRON
A halogen reacts to fill the gap in its outer shell to become more stable.

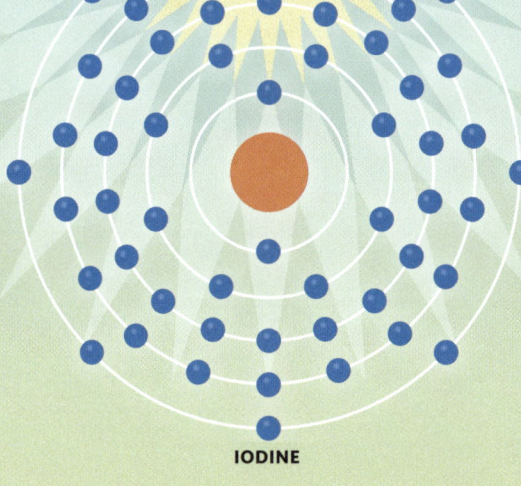

IODINE

ELECTRON GRABBERS

Halogens are nonmetals, each with one electron missing from their outer shell, which makes them highly reactive. These elements react with metals to complete their outer shell and produce salts. The reactivity of halogens decreases down the group in the periodic table – the atomic radius of each element increases, reducing the electron-grabbing forces of the nucleus.

32 | HALOGENS

HIGHLY VERSATILE

Extraordinary in their usefulness and flexibility, the transition metals occupy the centre of the periodic table and form the largest group of elements. They are also known as the d-block elements because their valence electrons gradually add to the d-orbitals (see p.11), filling them in sequence. Many transition metals form colourful ionic compounds that dissolve in water.

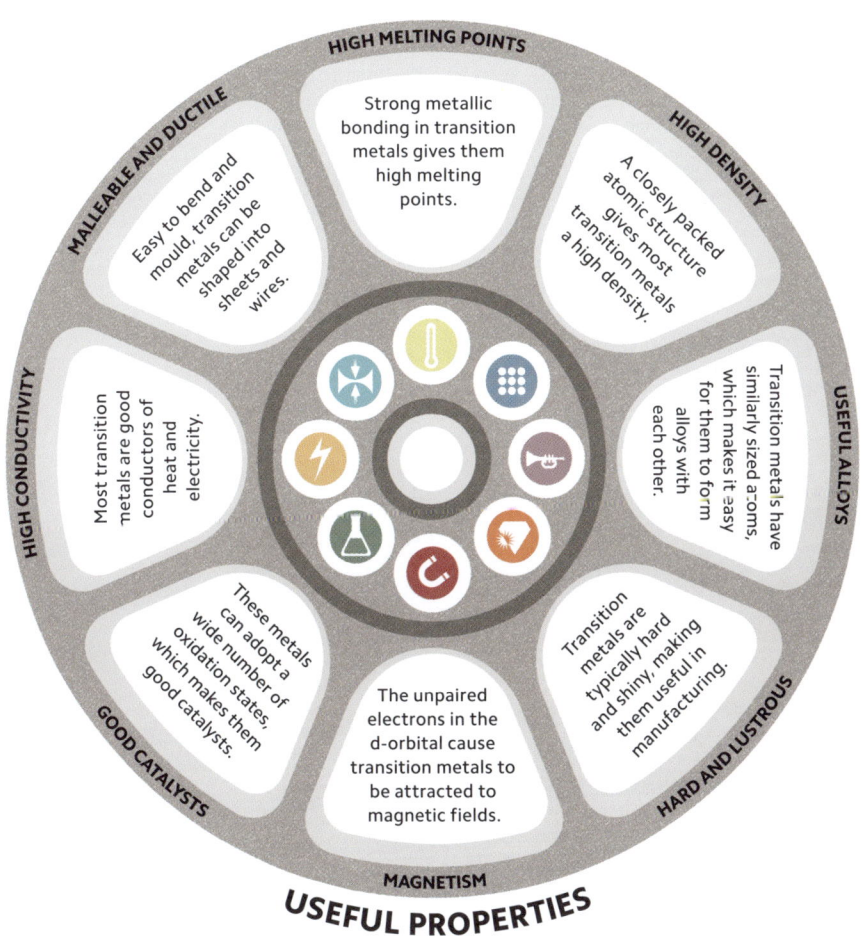

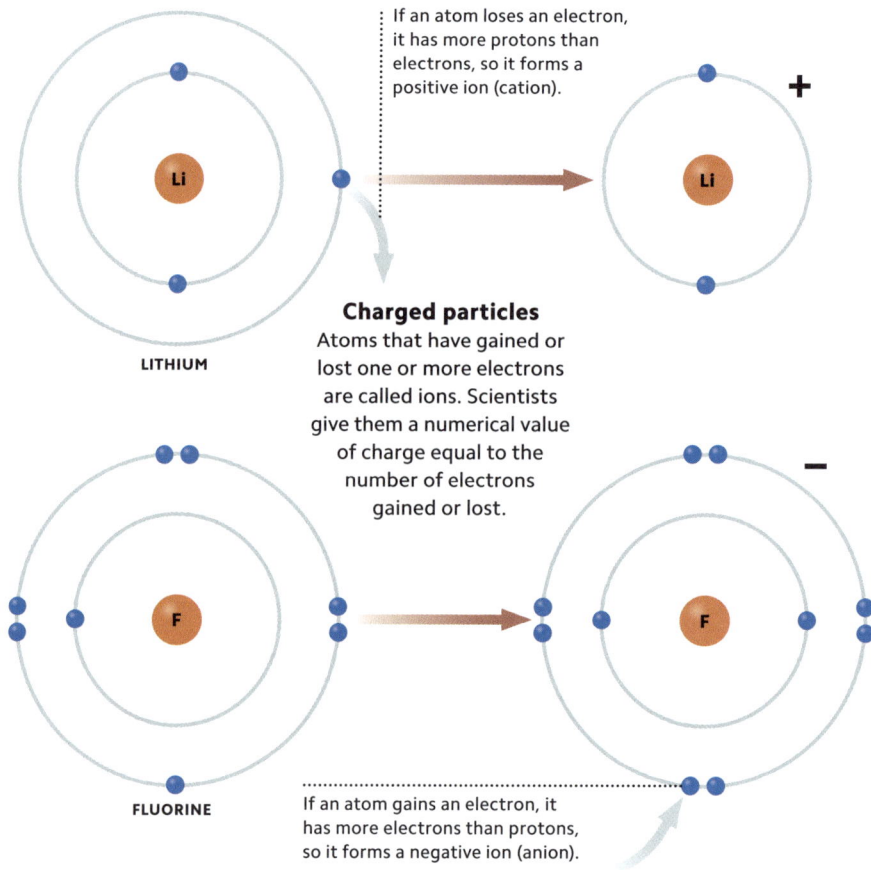

If an atom loses an electron, it has more protons than electrons, so it forms a positive ion (cation).

Charged particles
Atoms that have gained or lost one or more electrons are called ions. Scientists give them a numerical value of charge equal to the number of electrons gained or lost.

LITHIUM

FLUORINE

If an atom gains an electron, it has more electrons than protons, so it forms a negative ion (anion).

WIN OR LOSE

Atoms seek the stability of a full outer shell of electrons. To this end, they lose or gain valence electrons, becoming ions with an electric charge. Metals have few electrons in their outer shell and are able to lose them easily, forming positive ions called cations. Some nonmetals gain electrons to fill their shells, becoming negative ions called anions. This is key to how elements bond (see pp.36–37).

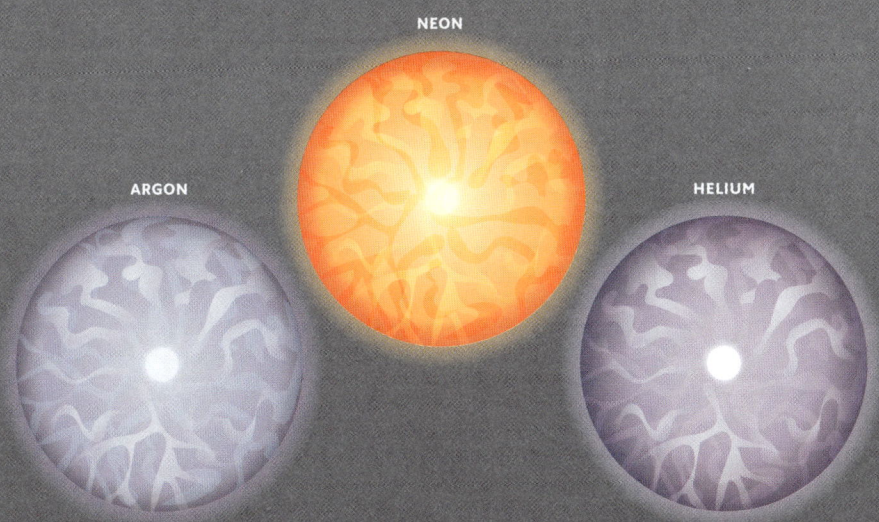

ARGON NEON HELIUM

UNTOUCHABLE ELEMENTS

Colourless, odourless, and inert, noble gases form the last column of the periodic table. They do not often react chemically because they have full outer electron shells, making them extremely stable. This group of elements were undiscovered for many years – the earliest periodic tables did not include them at all. However, once argon was isolated in 1894, scientists soon identified other members of the family.

KRYPTON

XENON

Noble gases
As most of the noble gases are colourless, they are visible only if they are trapped in glass spheres or bottles and then electrified.

ATOMS ASSEMBLE!

The interactions between an atom's valence electrons form different chemical bonds. When two atoms share a pair of electrons in order to fill their outer shells, a covalent bond is formed. In an ionic bond, a metal atom loses valence electrons and a nonmetal gains them. The electromagnetic attraction that results bonds the ions together.

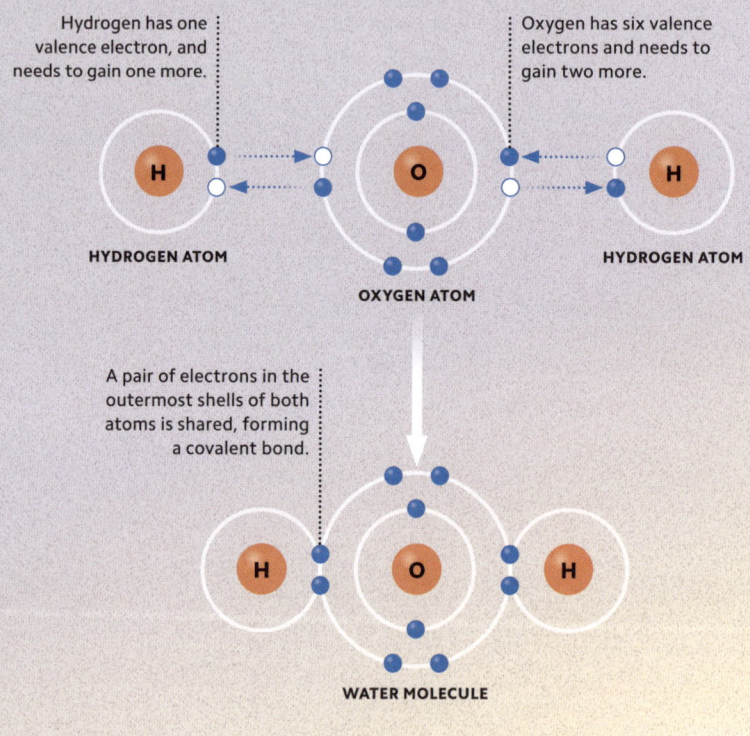

Hydrogen has one valence electron, and needs to gain one more.

Oxygen has six valence electrons and needs to gain two more.

HYDROGEN ATOM

OXYGEN ATOM

HYDROGEN ATOM

A pair of electrons in the outermost shells of both atoms is shared, forming a covalent bond.

WATER MOLECULE

Covalent bond
Covalent bonds form between atoms that have moderate electronegativity (see p.40) and do not give up their electrons easily.

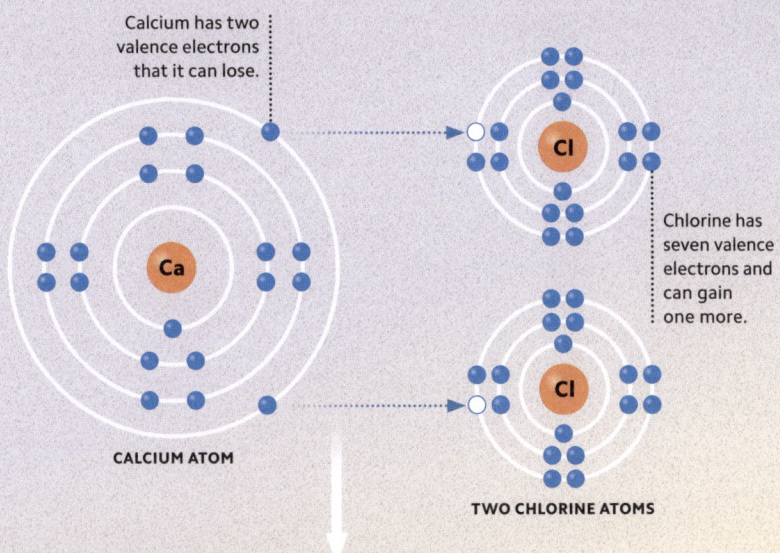

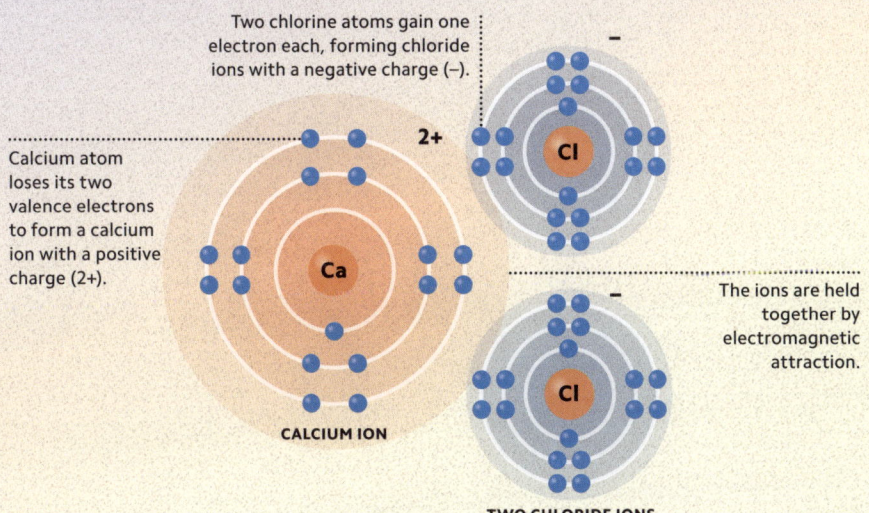

Ionic bond
Ionic bonds form between atoms with widely differing electronegativity. Metals can easily lose electrons, while nonmetals hold them tightly.

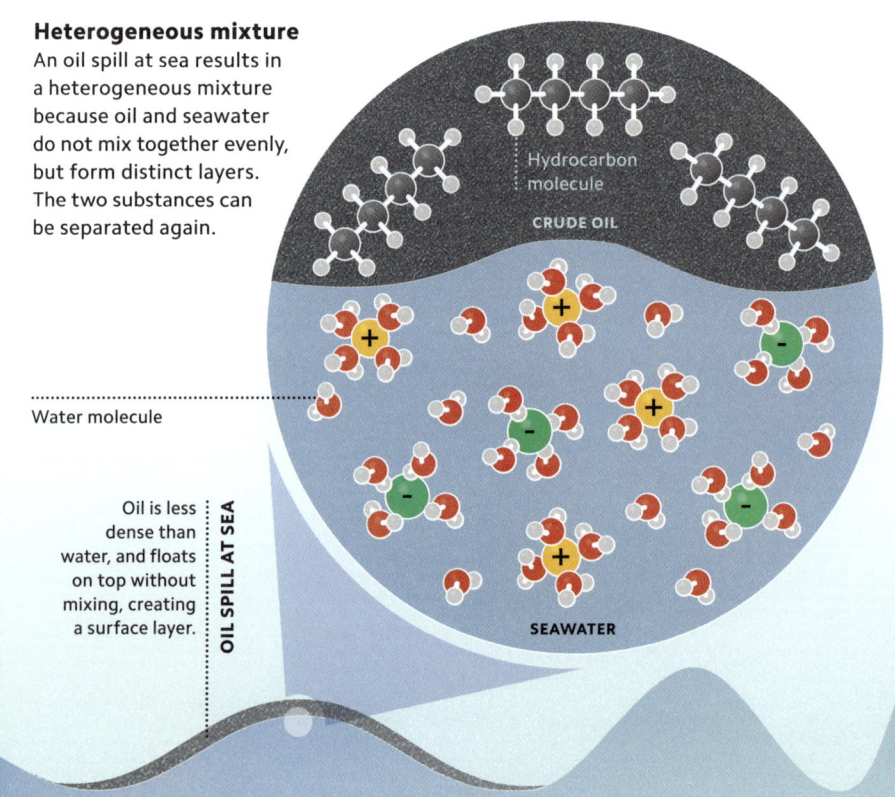

Heterogeneous mixture
An oil spill at sea results in a heterogeneous mixture because oil and seawater do not mix together evenly, but form distinct layers. The two substances can be separated again.

Hydrocarbon molecule

CRUDE OIL

Water molecule

Oil is less dense than water, and floats on top without mixing, creating a surface layer.

OIL SPILL AT SEA

SEAWATER

MOLECULAR BLENDS IN ACTION

A mixture is a combination of substances that do not react, and are not bonded together. The components retain their own properties and can be separated using a physical process, such as filtration or distillation. A solution is a homogeneous mixture in which particles of different substances are evenly distributed. The substances in a heterogeneous mixture are unevenly distributed and visually distinct. A suspension is a type of heterogeneous mixture in which solid particles are dispersed in a liquid or gas, but do not dissolve.

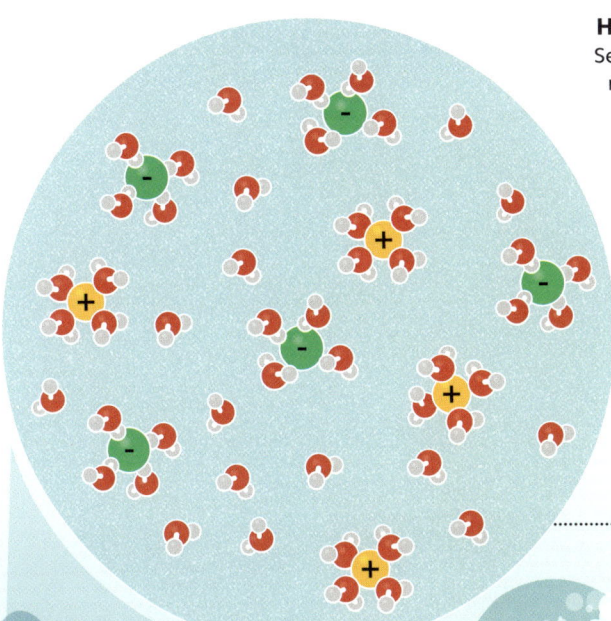

Homogeneous mixture
Seawater is a homogeneous mixture of salt and water. Dissolved sodium ions and chloride ions are surrounded by water molecules and are uniformly distributed. Evaporation can separate the components.

Seawater contains varying amounts of salt, but it is spread evenly throughout the mixture.
SEAWATER

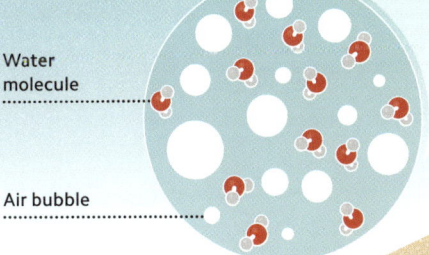

Water molecule

Air bubble

Sea foam is a heterogeneous mixture of water and air bubbles of different sizes.
FOAM

Sand grain

Wet sand is a suspension. Sand grains are dispersed unevenly in water and remain visually distinct.
WET SAND

SOLUTIONS AND OTHER MIXTURES | 39

ELECTRONIC TUG OF WAR

Electronegativity is a measure of how strongly an atom in a molecule pulls electrons towards itself. The higher the electronegativity of an atom, the more strongly it attracts shared electrons. Electronegativity influences how atoms combine, and in what ways they are likely to bond. In the periodic table (see pp.14–15), electronegativity increases left to right across periods and decreases going down each group.

Changes in attraction
Electronegativity increases across periods as the number of protons rises, increasing the electromagnetic pull on electrons. It decreases down each group, as atom size increases, with the pull of the nucleus spread across more electron shells.

Valence electrons are furthest from the nucleus, and shielded by inner shells in larger atoms, which decreases electronegativity.
SHIELDING

As the number of protons increases, so does the electronegativity.

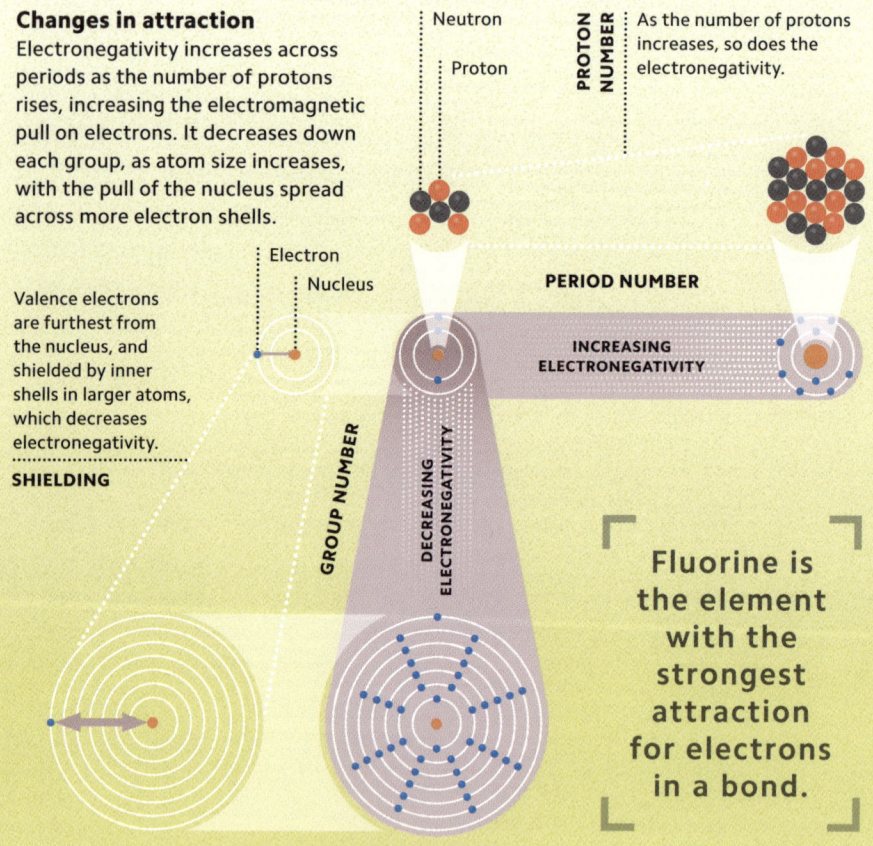

> Fluorine is the element with the strongest attraction for electrons in a bond.

The polarity of water
Water has two dipoles caused by unevenly distributed electrical charge. Each dipole is a tiny magnet, with a positive end and a negative end.

The oxygen atom has a negative partial charge.

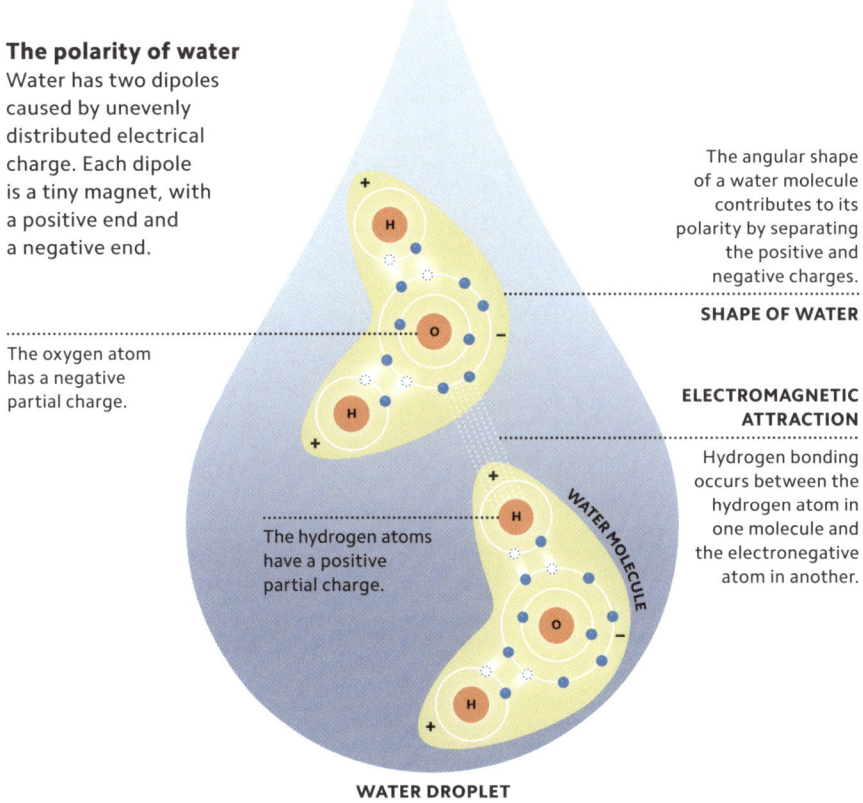

The angular shape of a water molecule contributes to its polarity by separating the positive and negative charges.

SHAPE OF WATER

ELECTROMAGNETIC ATTRACTION

Hydrogen bonding occurs between the hydrogen atom in one molecule and the electronegative atom in another.

The hydrogen atoms have a positive partial charge.

WATER DROPLET

MOLECULAR MAGNETS

The world would be a very different place without hydrogen bonds. These weak bonds arise between molecules in which a hydrogen atom is covalently bonded to a very electronegative atom, for example nitrogen or fluorine. The difference in electronegativity creates a dipole – two separate areas, one with positive and one with negative electric charge. The electromagnetic attraction between similar dipoles has a vital effect on the properties of water, and in many biological processes.

HYDROGEN BONDS

ATTRACTION EVERYWHERE

Weak attraction
A weak, short-range attraction between atoms or molecules arises from an uneven distribution of charge caused by electron cloud fluctuations.

SHIFTING CLOUDS
As the electron cloud in one atom or molecule shifts, it induces temporary positive and negative magnetic poles in other atoms or molecules.

Van der Waals forces are tiny attractions between atoms or molecules that arise from transient shifts in electron density. Electrons are often considered tiny particles, but in reality, they are more like clouds of electrical charge (see p.11). As a result of fluctuations in the cloud, a short-range net electromagnetic attraction comes into effect, impacting on the behaviour of solids, liquids, and gases.

ELECTRON DISTRIBUTION

TEMPORARY DIPOLE
If the electrons move to one side of a molecule, that side has a slight negative charge. The opposite side becomes positive.

ATTRACTIVE FORCE

VAN DER WAALS FORCES

THE LEAGUE OF ELEMENTS

Metals have lots of useful applications – if you know how they will behave. The reactivity series is a practical guide to the vigour of everyday metals' reactions to water or dilute acids. It shows whether a metal can displace another metal in a compound, which can also help determine how to extract a metal from its ore.

REACTION WITH WATER		REACTION WITH ACID	Element
VIOLENT FIZZING, HYDROGEN GAS AND ALKALINE METAL HYDROXIDE SOLUTION PRODUCED	MOST REACTIVE	EXPLOSIVE!	POTASSIUM (K)
			SODIUM (Na)
			LITHIUM (Li)
VERY SLOW REACTION		FIZZING, HYDROGEN GAS AND A SALT PRODUCED	CALCIUM (Ca)
			MAGNESIUM (Mg)
			ALUMINIUM (Al)
			ZINC (Zn)
			IRON (Fe)
SLIGHT REACTION WITH STEAM		REACTS SLOWLY WITH WARM ACID	TIN (Sn)
			LEAD (Pb)
NO REACTION	LEAST REACTIVE	NO REACTION	COPPER (Cu)
			SILVER (Ag)
			GOLD (Au)
			PLATINUM (Pt)

REACTIVITY SERIES | 43

LEFT- AND RIGHT-HANDED MOLECULES

Stereochemistry is the study of the 3D structural arrangement of the atoms in a molecule. Molecules with the same chemical formula but a different structure are called isomers. Some isomers are mirror images of each other, known as enantiomers. Even though they are seemingly so similar, the behaviour of enantiomers can be very different. In the case of limonene, it smells different depending on the enantiomer.

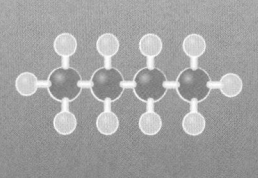
BUTANE

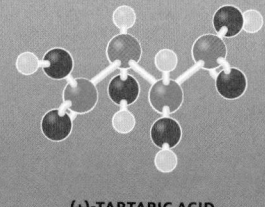
(+)-TARTARIC ACID

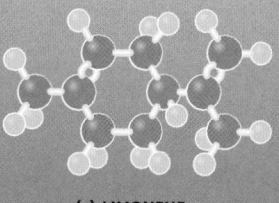
(+)-LIMONENE

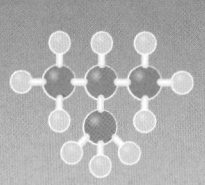
2-METHYLPROPANE

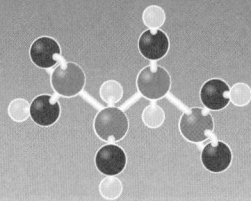
MESO-TARTARIC ACID

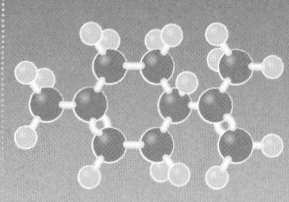
(-)-LIMONENE

Structural isomers
Structural isomers have the same formula but are connected differently, as with butane, a straight chain, and 2-methylpropane, a branched chain.

Diastereomers
Diastereomers have the same formula and the same connections, but the arrangements are different. In two forms of tartaric acid, this results in different solubility.

Enantiomers
Enantiomers, or optical isomers, have the same formula, but the structures are mirror images that cannot be superimposed over each other, like a pair of hands.

FACE-CENTRED CUBIC

Table salt (NaCl) has an ion at each corner and the same type of ion at the centre of each face. Diamond has carbon atoms arranged in two superimposed face-centred cubes.

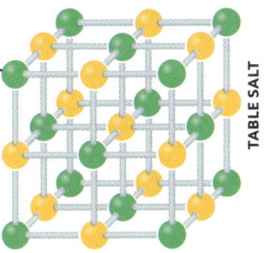

TABLE SALT

This structure, with one atom at each corner of the cube and one in the centre, is largely found in metals, including chromium, iron, and tungsten.

BODY-CENTRED CUBIC

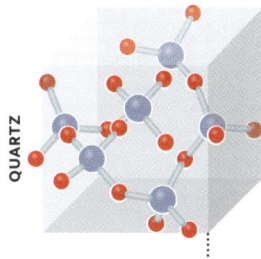

QUARTZ

Ordered structure
Unit cells are the building blocks of the 3D pattern of the entire crystal lattice.

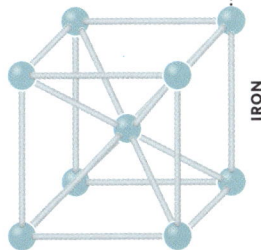

IRON

In quartz (SiO_2) each silicon atom (blue) is bonded to four oxygen atoms (red), in a tetrahedral arrangement, which forms a continuous lattice.

TRIGONAL

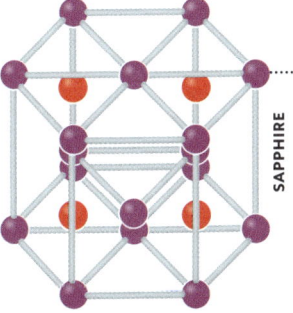

SAPPHIRE

HEXAGONAL CLOSE-PACKED

Sapphire (Al_2O_3) consists of oxygen ions (red) and aluminium ions (purple) forming closely packed layers, arranged in a six-sided (hexagonal) pattern.

REGULAR BEAUTIES

Crystalline solids have a highly ordered, repeating structure, called a lattice. Rocks, minerals, and metals are all typically crystalline. The regular lattice arrangement of their atoms determines properties such as hardness, melting point, conductivity, and – when it comes to flawless precious gems – value. Crystal structure is described via the smallest possible group of particles that repeats, called the unit cell.

CRYSTALS AND GEMS

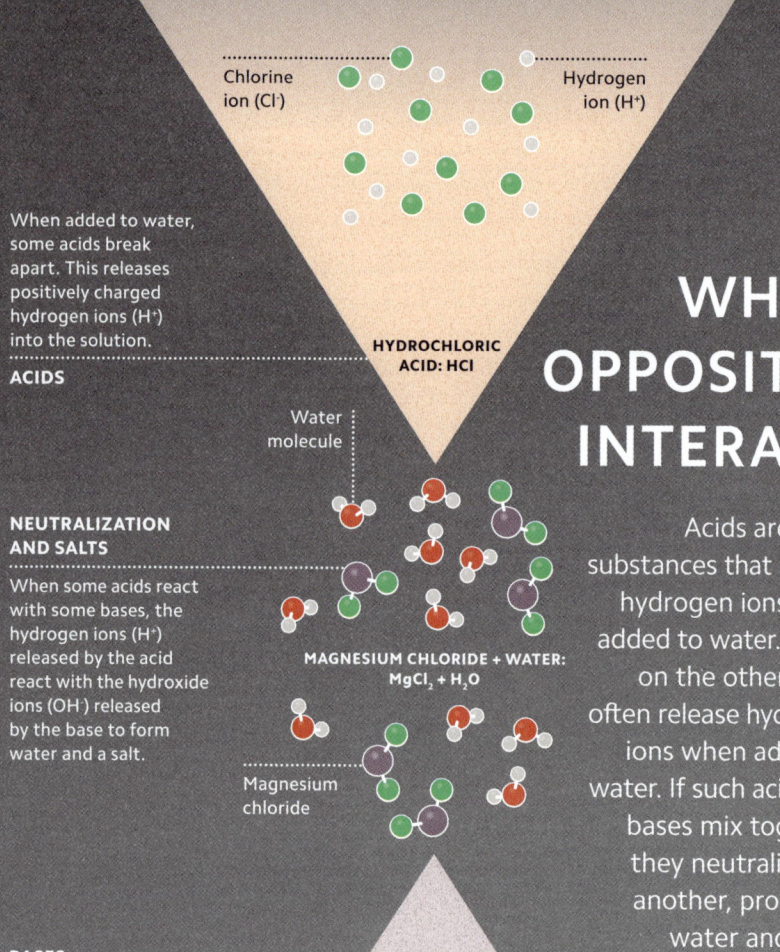

WHEN OPPOSITES INTERACT

Acids are often substances that release hydrogen ions when added to water. Bases, on the other hand, often release hydroxide ions when added to water. If such acids and bases mix together, they neutralize one another, producing water and a salt (a compound of positively and negatively charged ions).

ACIDS

When added to water, some acids break apart. This releases positively charged hydrogen ions (H^+) into the solution.

NEUTRALIZATION AND SALTS

When some acids react with some bases, the hydrogen ions (H^+) released by the acid react with the hydroxide ions (OH^-) released by the base to form water and a salt.

BASES

When added to water, some bases release hydroxide ions, which readily lose electrons to other atoms. A base that dissolves in water is called an alkali.

46   ACIDS, BASES, AND SALTS

MEASURING ACIDITY

Acids and bases are ranked on the pH scale, which is a measure of the concentration of hydrogen ions in a solution. The higher the concentration, the more acidic a substance. The pH value of a substance can be measured using universal indicator (see pp.82–83), which changes colour based on the pH level. An electronic probe can also measure concentration of hydrogen ions in a solution.

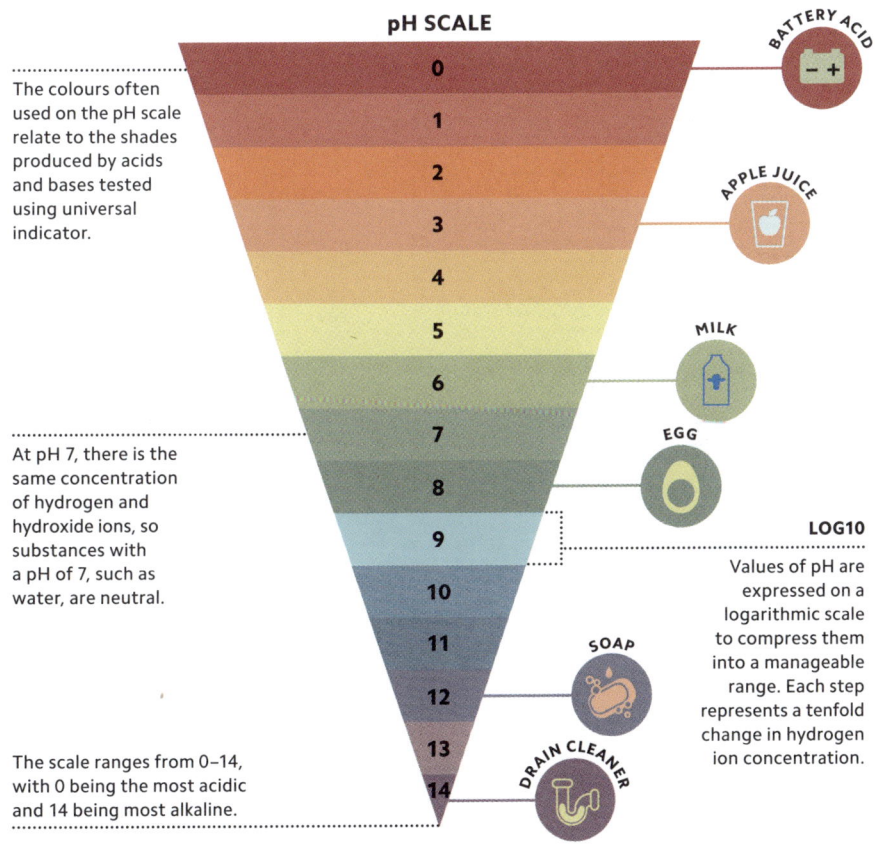

The colours often used on the pH scale relate to the shades produced by acids and bases tested using universal indicator.

At pH 7, there is the same concentration of hydrogen and hydroxide ions, so substances with a pH of 7, such as water, are neutral.

The scale ranges from 0–14, with 0 being the most acidic and 14 being most alkaline.

LOG10

Values of pH are expressed on a logarithmic scale to compress them into a manageable range. Each step represents a tenfold change in hydrogen ion concentration.

THE PH SCALE | 47

Aromatic molecules, or arenes, have a cyclic structure with atoms joined by electrons shared in especially strong covalent bonds. Benzene (C_6H_6) is the best-known example. It is an organic molecule consisting of a ring of six carbon atoms, each attached to one hydrogen atom. Electrons in the carbon atoms are delocalized, meaning they are shared evenly around the ring.

CYCLIC STABILITY

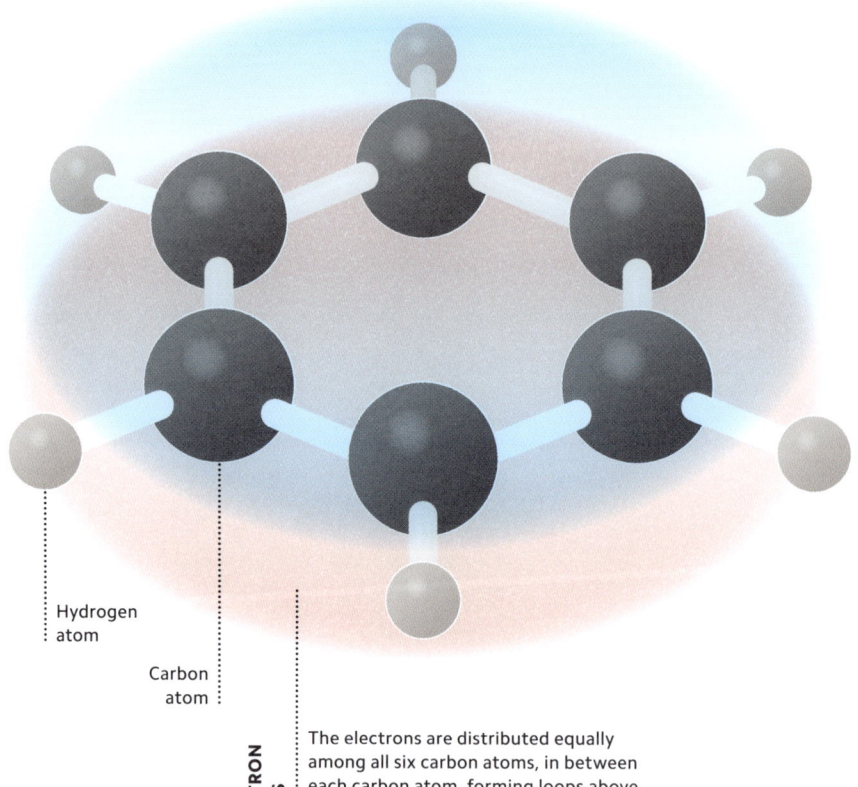

Hydrogen atom

Carbon atom

ELECTRON RINGS: The electrons are distributed equally among all six carbon atoms, in between each carbon atom, forming loops above and below them. This gives benzene excellent stability.

48 | AROMATIC MOLECULES

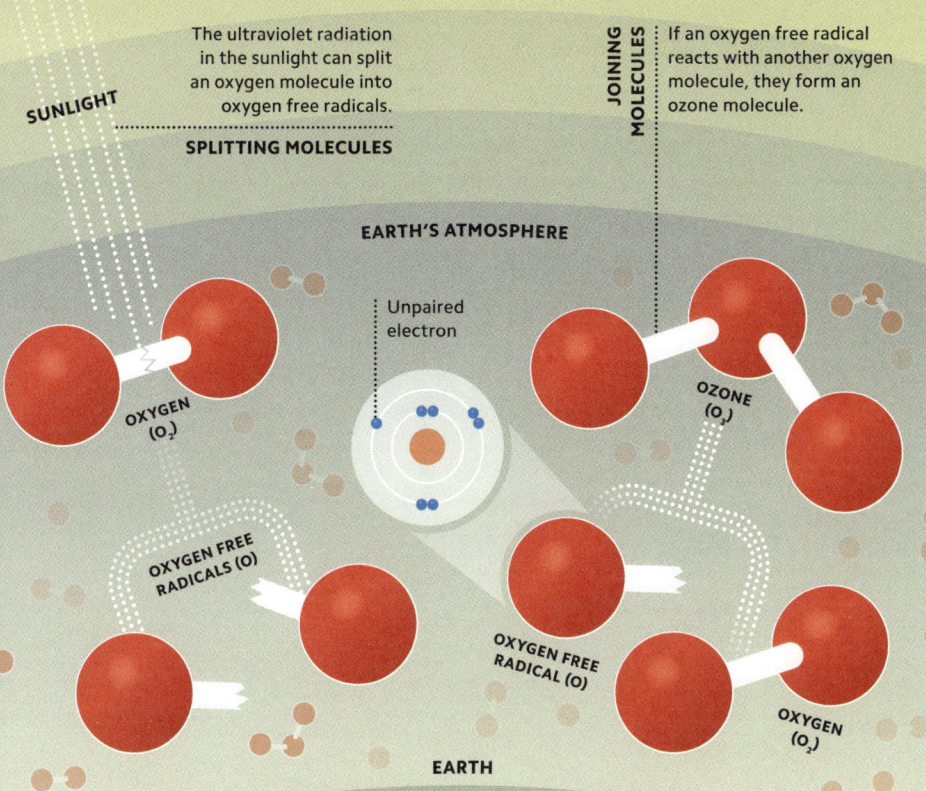

GREEDY ATOMS

A free radical is an atom, molecule, or ion with an unpaired electron left dangling after a bond breaks. Radical formation can result from the effects of ionizing radiation (high-energy waves or particles that can free an electron from an atom), heat, light, or electrical or chemical processes. For example, when an oxygen molecule in the atmosphere is split by ultraviolet radiation, oxygen free radicals are produced, which react further to form ozone. Free radicals are highly reactive towards other substances as well as each other. In order to pair their lone electrons, they can sometimes trigger chain reactions.

CHEMI
PROCE

CHEMISTRY
PROCESSES

From combustion to corrosion to cookery, chemistry drives the processes by which we live and shape our world. In any chemical reaction, such as addition, elimination, or substitution, the atoms in the relevant compounds rearrange through the breaking and reforming of bonds. Such transformations involve the absorption and release of heat, as Antoine Lavoisier discovered in the 18th century. This laid the groundwork for modern chemistry to develop, with its understanding that matter behaves in quantitative and predictable ways. Today, chemists can precisely refine their processes to achieve maximum yield and efficiency, with improved consistency and sustainability.

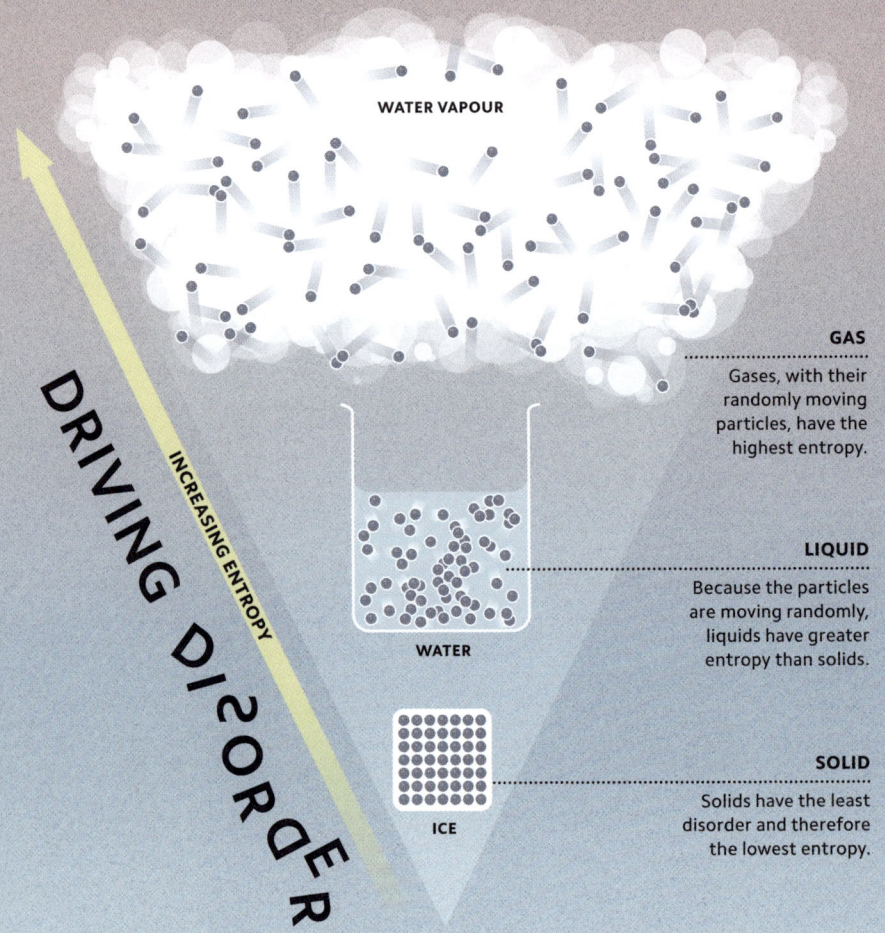

DRIVING DISORDER

INCREASING ENTROPY

WATER VAPOUR

GAS
Gases, with their randomly moving particles, have the highest entropy.

WATER

LIQUID
Because the particles are moving randomly, liquids have greater entropy than solids.

ICE

SOLID
Solids have the least disorder and therefore the lowest entropy.

Entropy can be thought of as the measure of disorder within a system. Systems tend to naturally become more random over time because it takes more energy to keep them organized. Imagine dropping black ink into water. The ink and water would gradually mix, and it would be difficult to separate the two substances. Different substances in different forms have more or less entropy. For example, ice, water, and water vapour are all H_2O, but liquid water has greater entropy than the same number of molecules in ice.

Breaking bonds
Bond enthalpy, the energy needed to break chemical bonds, is defined as the energy needed to break one mole (see p.22) of bonds to give separated atoms.

Bond-breaking is endothermic – it requires the addition of energy to break bonds.

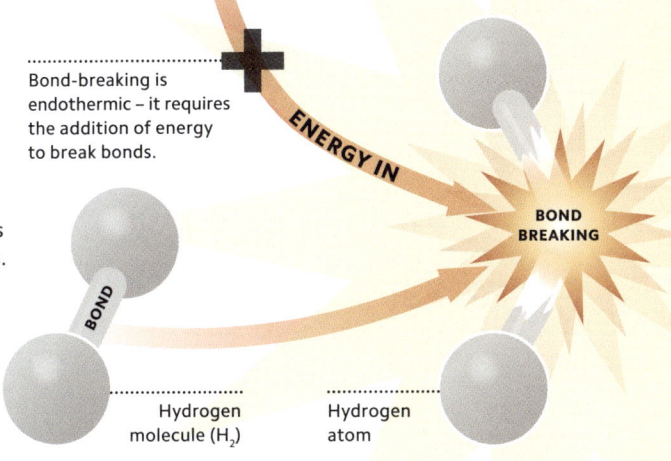

BOND ENERGIES

Enthalpy is the total heat content of a system. When compounds react, bonds have to break. The amount of energy this takes for covalent compounds (see p.36) in a gas is called the bond enthalpy, and it is a positive number. The same amount of energy is released when a new bond of the same type forms, but it is a negative number. The making and breaking of several bonds happen in a chemical reaction, so overall energy change, also called enthalpy, can be positive or negative.

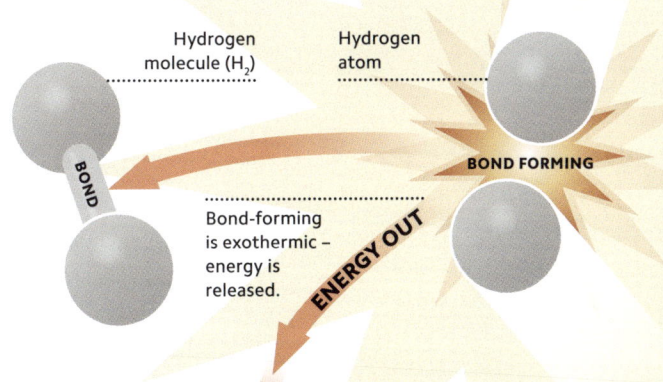

Bond-forming is exothermic – energy is released.

Forming bonds
Free atoms are at a higher energy state than stable, bonded ones, so forming chemical bonds gives out energy – the same amount as it originally took to break them.

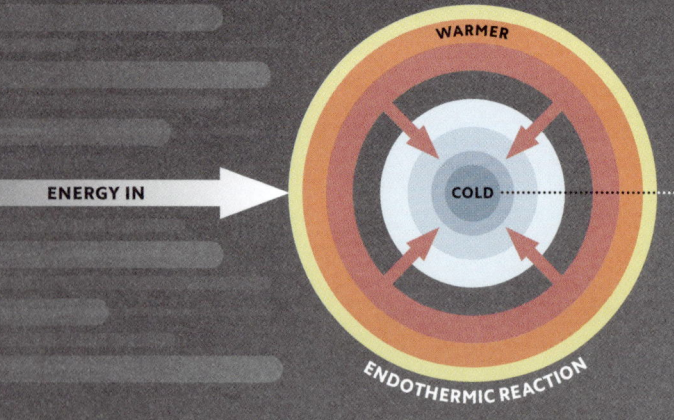

COOLING SURROUNDINGS

Endothermic reactions take energy from their surroundings. Since we are part of the surroundings, these reactions often feel cool.

ENERGY IN AND OUT

When chemical reactions occur, there is a change in energy. The balance – between the energy needed to break bonds and the energy released when new bonds form – can be either negative or positive (see p.53). That is, energy is given out overall, or taken in overall. Chemical reactions that give out energy make their surroundings feel warm, and are called exothermic. Reactions that take in energy tend to make their surroundings feel cooler, and are called endothermic.

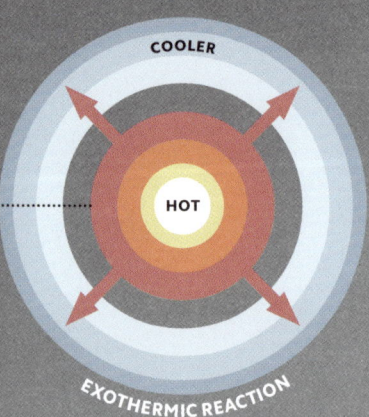

HEATING SURROUNDINGS

Exothermic reactions give out energy to their surroundings. To us, they often feel hot. The substances actually reacting, however, end up in a lower energy state.

HYDROGEN MOLECULE: Two hydrogen molecules each have a single bond. There are four hydrogen atoms in total.

OXYGEN MOLECULE: A single oxygen molecule contains a double bond between two oxygen atoms.

WATER MOLECULES: Two water molecules are made: in each one oxygen atom is joined to two hydrogen atoms by two single bonds.

$$2H_2 + O_2 \longrightarrow 2H_2O$$

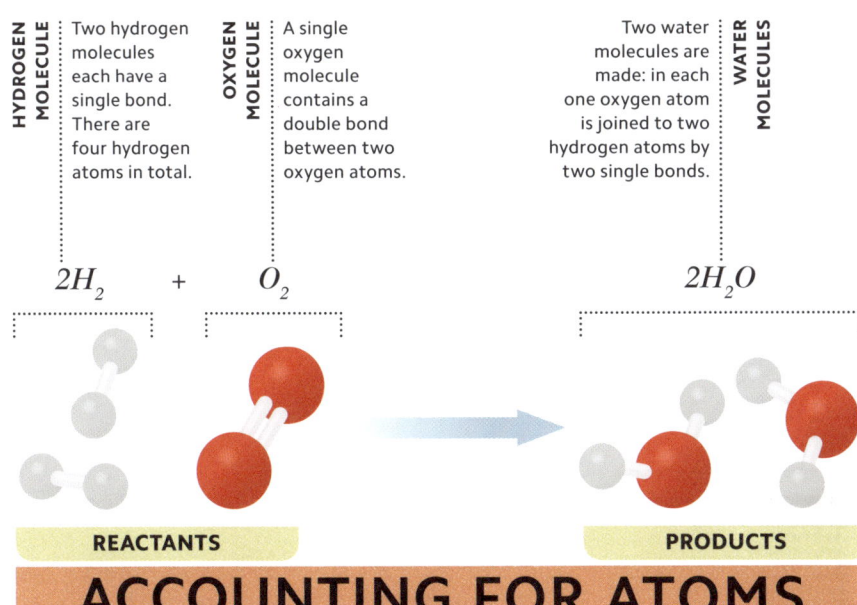

REACTANTS → PRODUCTS

ACCOUNTING FOR ATOMS

MASS IS BALANCED: There are the same numbers and types of atoms on both sides of the arrow. The scale is "balanced".

Whenever a chemical reaction happens, atoms move around, but no further atoms can be created or destroyed. For example, when hydrogen and oxygen react to form water, there are four hydrogen atoms and two oxygen atoms before and after the reaction. The bonds have changed and the atoms have rearranged, but the total number and type of atoms has stayed the same throughout: mass is conserved in all chemical reactions.

SWAPPING PARTNERS

In chemical reactions, the chemical bonds between atoms break and form. Bonds in the reactants (the substances at the start of a reaction) break, while new, different bonds form in the products (the substances resulting from the reaction). Energy is needed to break bonds, and it is released when they form. Different types of bond absorb and release different amounts of energy – for example, the triple bond in N_2 molecules is very strong, absorbing and releasing a lot of energy.

Breaking and making bonds
Chemical reactions involve bond-breaking, which requires energy, and bond-forming, which releases it. Bonds break in the reactants and form in the products.

NITROGEN AND HYDROGEN
Three hydrogen molecules (H_2) are needed to react with one nitrogen molecule (N_2).

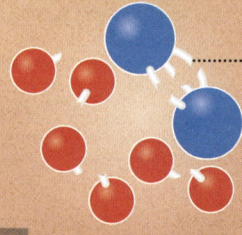

BONDS BREAK
First, the bonds in the reactants (H_2 and N_2) have to break. This requires energy, such as heat or pressure.

> Nitrogen molecules make up 78% of Earth's atmosphere and have one of the strongest bonds.

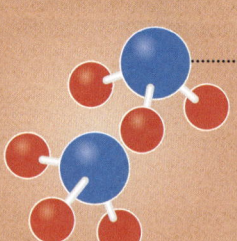

NEW BONDS FORM
Bonds form between the nitrogen and hydrogen to form ammonia (NH_3). This releases energy.

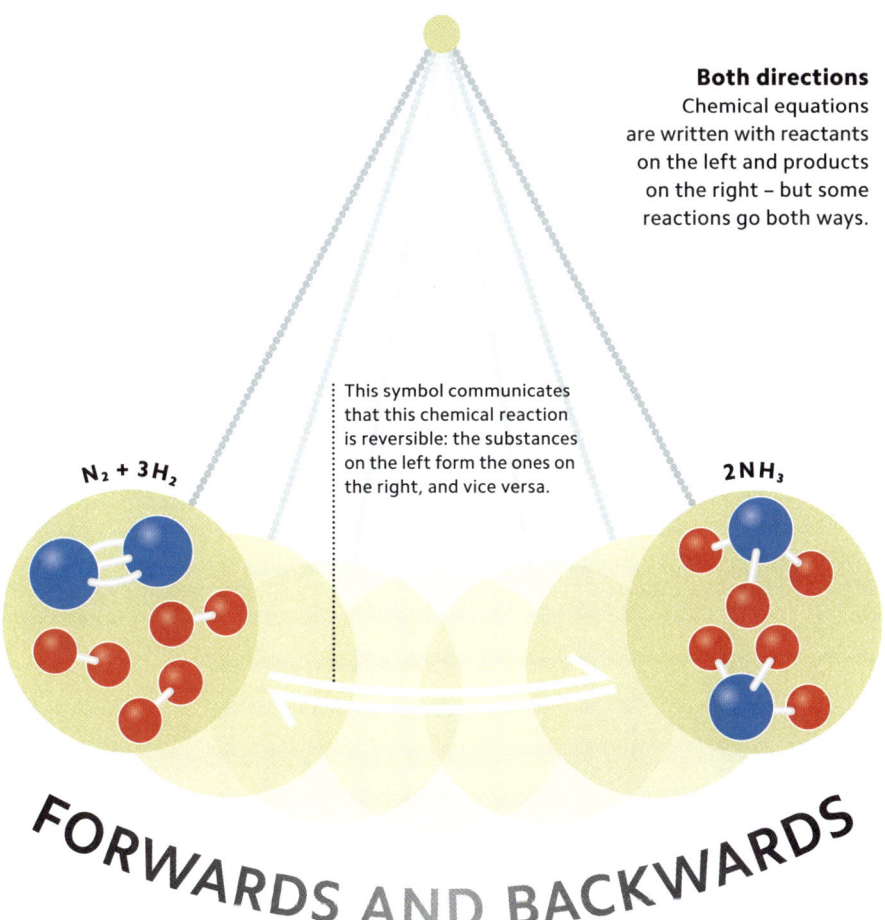

Both directions
Chemical equations are written with reactants on the left and products on the right – but some reactions go both ways.

This symbol communicates that this chemical reaction is reversible: the substances on the left form the ones on the right, and vice versa.

$N_2 + 3H_2$

$2NH_3$

FORWARDS AND BACKWARDS

We often think of chemical reactions as going in one direction but, depending on the conditions (see pp.62–63), what we think of as the products can sometimes react with each other to form the reactants. Such reactions are described as reversible. For example, the reaction between nitrogen (N_2) and hydrogen (H_2) to make ammonia (NH_3) is reversible. At high pressures, because there are fewer individual molecules of the product, more ammonia is produced. Under other conditions, however, ammonia is more likely to be converted into nitrogen and hydrogen (the reactants).

REVERSIBLE REACTIONS | 57

Gibbs free energy (G) is a measure of how much energy is available to do work in a system. The formula below combines enthalpy, entropy, and temperature, and helps to predict whether a chemical reaction will happen at a constant temperature and pressure. A negative change in Gibbs free energy where G decreases overall means that the reaction can happen spontaneously without additional energy. A positive change means the reaction will not occur. Some reactions only happen spontaneously above, or below, specific temperatures, and some reactions are not feasible at any temperature.

WILL REACTIONS HAPPEN?

Calculating ΔG
The formula below is used to calculate a change in Gibbs free energy.

CHANGE IN ENTHALPY
This is the heat absorbed or released during the chemical reaction.

CHANGE IN ENTROPY
This represents the change in entropy (see p.52) during the chemical reaction.

$$\Delta G = \Delta H - T \times \Delta S$$

CHANGE IN GIBBS FREE ENERGY
The Δ symbol means change: the change in G during a chemical reaction represents the usable energy available.

TEMPERATURE (IN KELVIN)
The temperature at which the reaction takes place, expressed in Kelvin, a unit of measurement for temperature. This is not a change.

58 | GIBBS FREE ENERGY

Particles need to collide to react (see pp.60–61), but they need to collide with enough energy. This minimum amount of energy needed is called the activation energy. For an endothermic reaction (see p.54), the products end up at a higher energy than the reactants, so the enthalpy change, ΔH, is positive. For an exothermic reaction (see p.54) the reverse is true. Either way, there is always an energy barrier to get over first: reactions only proceed when particles have enough energy to exceed the activation energy.

CHEMISTRY'S INVISIBLE BARRIER

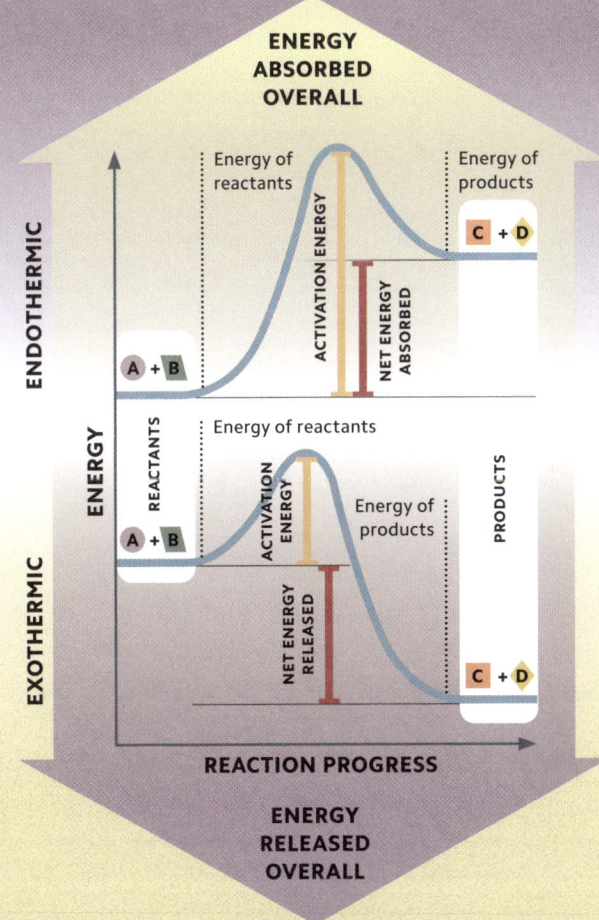

Activation energy
Energy must always be applied for a reaction to start, and this energy "barrier" is called the activation energy. Some reactions need very little energy to get started. If there is a high activation energy barrier, the reaction may never start.

ACTIVATION ENERGY | 59

Low probability
Particles have less chance of reacting with a smaller surface area, or at a lower temperature, pressure, or concentration.

SMALL SURFACE AREA
Solid reactants that have a small surface area are less likely to undergo chemical reactions.

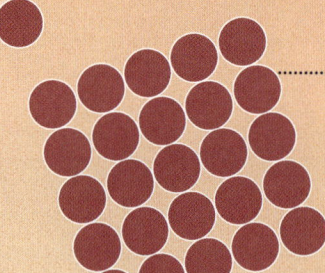

LOW PRESSURE AND CONCENTRATION
Fewer collisions between particles take place at low pressure and concentration, reducing the reaction rate.

INCORRECT ANGLE
Particles only react together if they collide at the right orientation, otherwise they bounce off.

LOW-ENERGY COLLISIONS
Particles at lower temperatures are less likely to collide with enough energy for them to react.

MAKING AN IMPACT

Chemical reactions take place when particles collide. But the particles need to impact with enough energy and in the right orientation for existing bonds to break and new ones to form. Chemists boost the energy and frequency of collisions by increasing the temperature, concentration, or surface area of the reactants, thus increasing the reaction rate. Catalysts (see p.70) and enzymes (see p.105) can also speed up a reaction.

High probability
Increasing the energy or surface area of molecules, or bringing them closer together, increases the likelihood of reactions occurring.

LARGE SURFACE AREA
Increasing the surface area of solid reactants, for example by breaking a big piece into smaller ones, increases the reaction rate, as more collisions will take place.

HIGH PRESSURE AND CONCENTRATION
There are more collisions when the reactants are at higher pressure or concentration.

HIGH-ENERGY COLLISIONS
At higher temperatures, reactants are more likely to collide with the energy required for them to react.

CORRECT ANGLE
The reactants must collide in an orientation that brings the right parts of the molecules into contact.

COLLISION THEORY

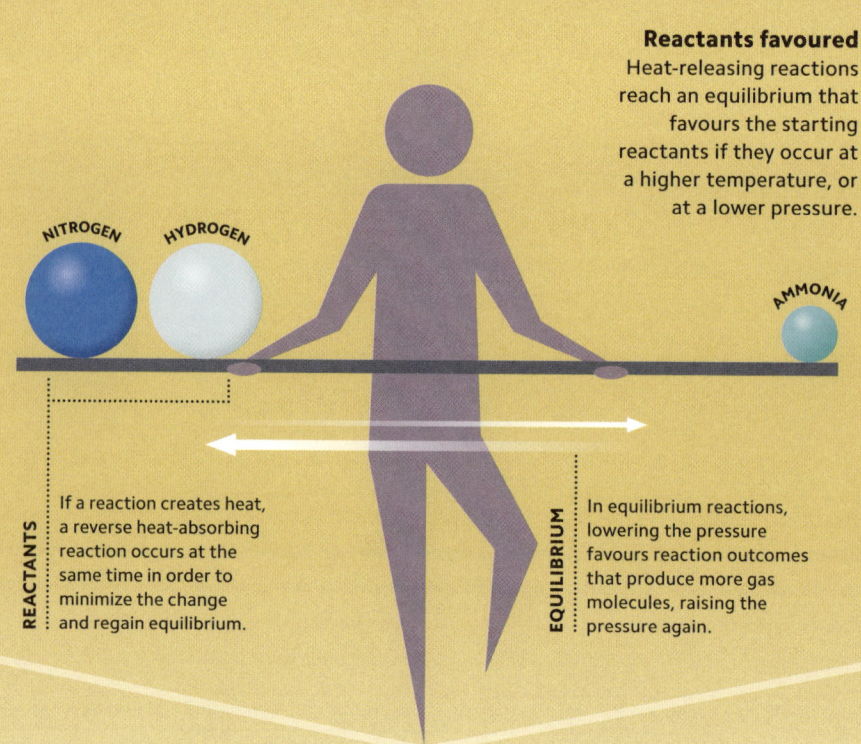

Reactants favoured
Heat-releasing reactions reach an equilibrium that favours the starting reactants if they occur at a higher temperature, or at a lower pressure.

REACTANTS: If a reaction creates heat, a reverse heat-absorbing reaction occurs at the same time in order to minimize the change and regain equilibrium.

EQUILIBRIUM: In equilibrium reactions, lowering the pressure favours reaction outcomes that produce more gas molecules, raising the pressure again.

FINDING BALANCE

Some chemical reactions are reversible and can go in both directions (see p.57). Reactants bond to form products, and at the same time, products turn back into reactants, eventually reaching a balance point called equilibrium. After this, even though the forward and backward reactions continue, the ratio of reactants to products does not change. If the conditions in a system are altered, the position of the equilibrium will shift to minimize the change, which is known as Le Chatelier's principle.

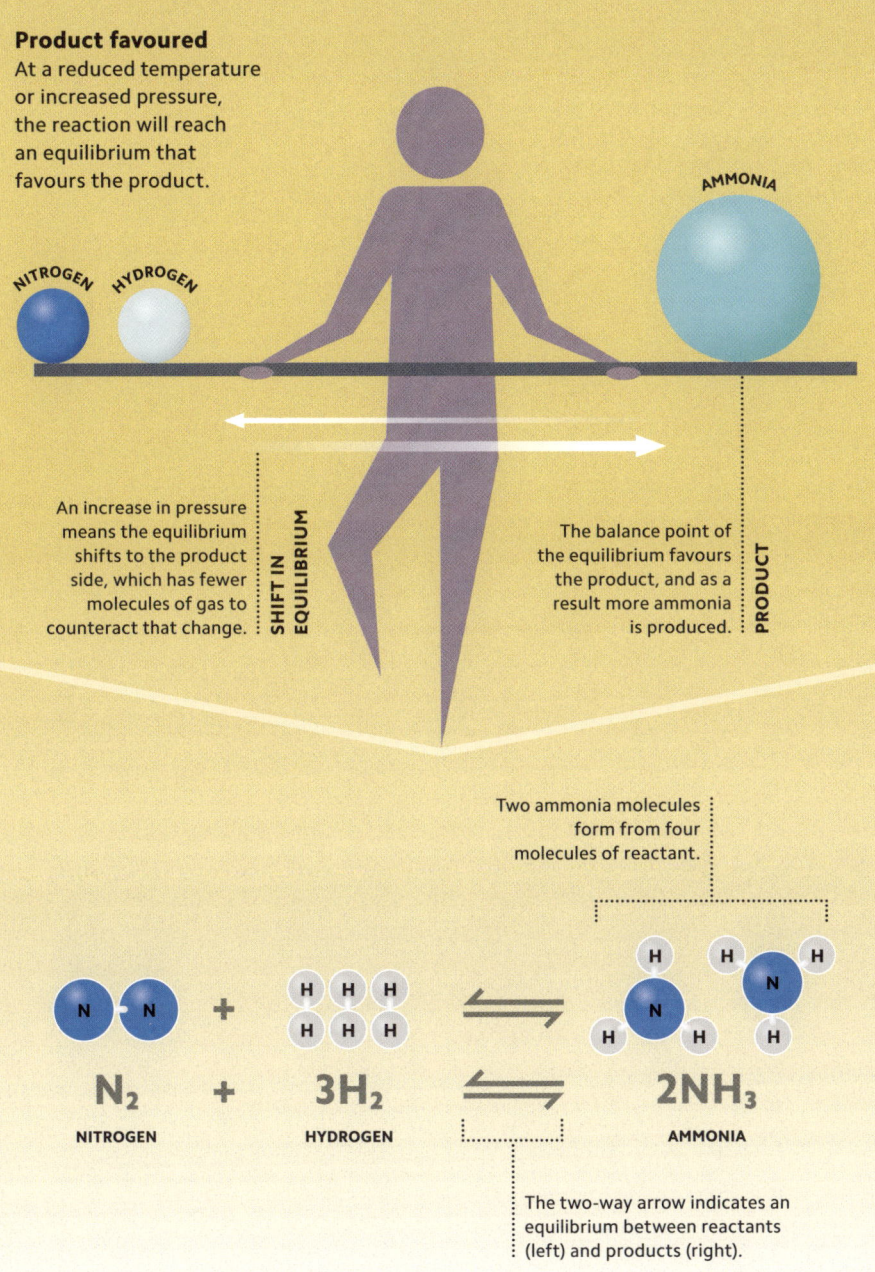

REACTION TYPES

CHEMICAL

In a chemical reaction, one or more substances (reactants) are converted into one or more new substances (products). In organic chemistry – the chemistry of carbon-based molecules – there are three main types of reaction. In addition reactions, one molecule is added to another to make a new, larger molecule and no other products. In elimination reactions, a small molecule is eliminated from a larger one, leaving two molecules behind. In substitution reactions, an atom, or group of atoms, is substituted with something else.

Addition reaction
Molecules with C=C bonds are often involved in addition reactions. The double bond breaks, allowing new bonds to form.

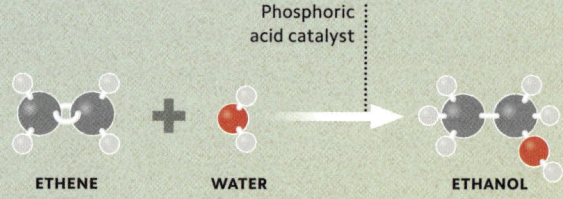

Elimination reaction
These occur when a smaller molecule is formed from a larger one, leaving two products behind.

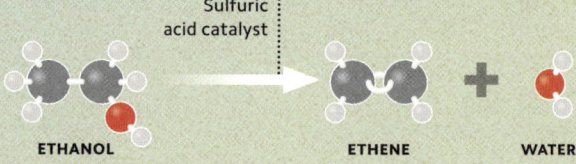

Substitution reaction
These reactions involve one atom, or group of atoms, being swapped for another. Here, NO_2 replaces an H atom.

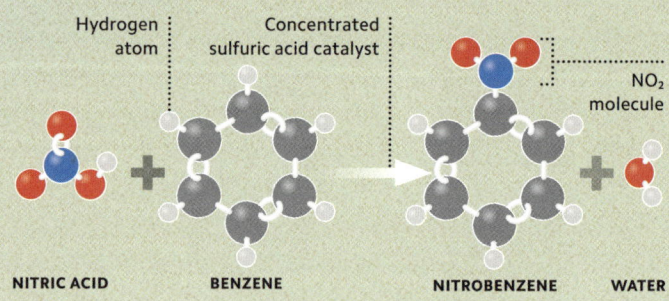

Musical chairs

Imagine a game of musical chairs for the idea of limiting reactants: once all the chairs are gone, no further "reaction" is possible.

The "product" is made by combining one chair and one person.

PRODUCT

PUSHING TO COMPLETION

This substance is "in excess" – it has nothing to react with.

EXCESS REACTANT

Often in a chemical reaction we do not have precisely the right amount of each of the reactants to react with each other. Usually, there is more of one than is needed, and this substance is described as being "in excess". The other reactant is said to be the "limiting reactant" because, once it has all reacted, the reaction stops. Any of the reactants left over are excess.

LIMITING REACTANTS | 65

OVERWHELMING OXYGEN

During a process called oxidation, most metals react with oxygen to form generally more stable oxides. In this process, electrons from the metal transfer to oxygen atoms. Many metals, if left exposed to air, corrode, deteriorating due to reactions with oxygen, water, and other substances in the environment. Rust is a form of corrosion that forms on iron when the metal loses electrons to form Fe^+ ions. Another example of corrosion is the green-blue patina called verdigris that forms on copper.

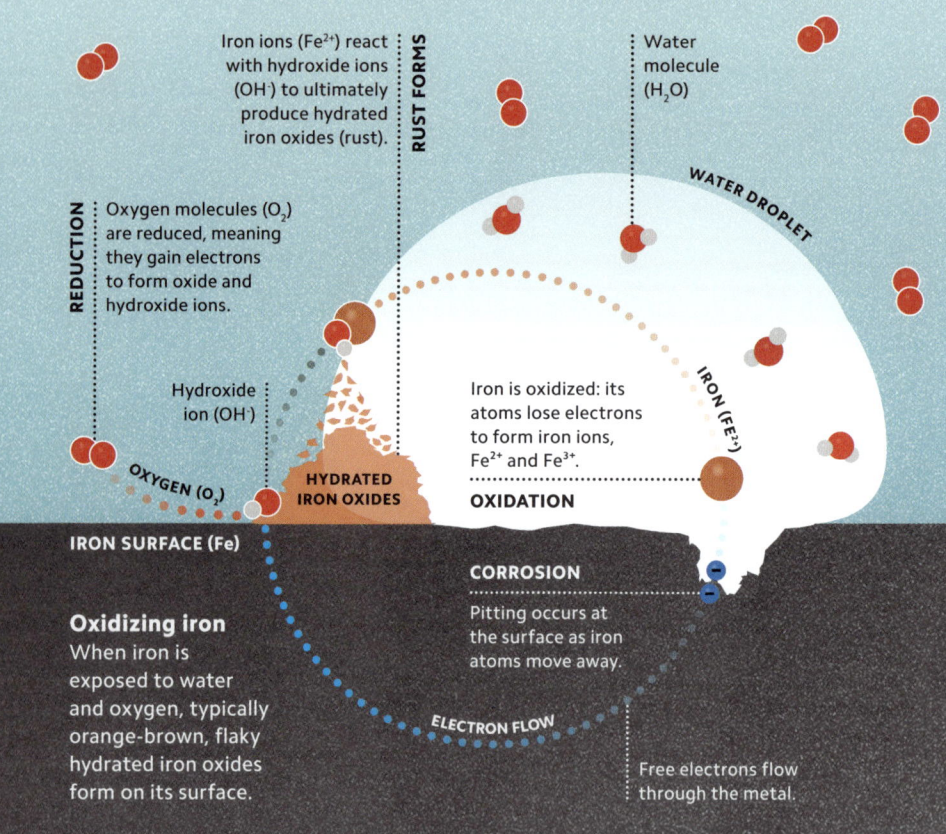

RUST FORMS
Iron ions (Fe^{2+}) react with hydroxide ions (OH^-) to ultimately produce hydrated iron oxides (rust).

Water molecule (H_2O)

REDUCTION
Oxygen molecules (O_2) are reduced, meaning they gain electrons to form oxide and hydroxide ions.

WATER DROPLET

Hydroxide ion (OH^-)

Iron is oxidized: its atoms lose electrons to form iron ions, Fe^{2+} and Fe^{3+}.

OXIDATION

IRON (FE³⁺)

OXYGEN (O₂)

HYDRATED IRON OXIDES

IRON SURFACE (Fe)

CORROSION
Pitting occurs at the surface as iron atoms move away.

Oxidizing iron
When iron is exposed to water and oxygen, typically orange-brown, flaky hydrated iron oxides form on its surface.

ELECTRON FLOW

Free electrons flow through the metal.

66 | OXIDATION AND CORROSION

MATERIALS IN

Iron ore (rocks containing iron oxides), coke (carbon), and limestone enter the furnace.

The hot waste gases are used to heat the air going in below, so little energy is wasted.

WASTE GASES

BLAST FURNACE

GETTING METAL

Metals are often not found in their pure form but rather combined with other elements, especially oxygen. Unless it can be recycled, iron (Fe) has to be extracted from iron ores, such as haematite (Fe_2O_3) and magnetite (Fe_3O_4). Adding carbon in the form of coke to a furnace forms carbon monoxide (CO). This acts as a reducing agent, "taking" the oxygen from the ore to produce iron (Fe) and carbon dioxide (CO_2). Refining the metal that forms, called pig iron, further makes iron's best-known alloy, steel.

HOT AIR BLAST

Hot air is blasted into the furnace to maintain a high temperature.

MOLTEN SLAG

MOLTEN IRON

REDUCTION

The iron ore has been reduced: iron ions have gained electrons to produce iron metal.

MOLTEN SLAG

The limestone reacts with rocky material from impurities in the iron ore to form calcium silicate, called slag.

REDUCTION FROM OXIDES | 67

FIRE, FURY, AND REACTIONS

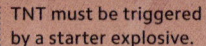

Explosive reaction
TNT is less sensitive than some other explosives and only blows up when it is detonated. However, when it does explode, it is extremely destructive.

DETONATION

TNT must be triggered by a starter explosive.

DECOMPOSITION

A decomposition reaction, such as TNT exploding, is one in which a single starting substance breaks apart to form two or more products.

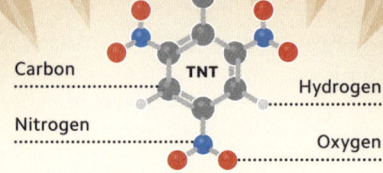

Carbon — TNT — Hydrogen
Nitrogen — Oxygen

Some chemical reactions release a lot of energy. During combustion, fuel reacts with oxygen to produce heat. Combustion needs oxygen, fuel, and heat to keep going. Remove any one of these and the fire goes out. Explosions are rapid reactions that release a huge amount of energy very quickly and with a destructive force. When TNT (2,4,6-trinitrotoluene) is detonated with heat or an electric charge, the molecule flies apart, forming nitrogen, hydrogen, carbon monoxide, and carbon.

$$2C_7H_5N_3O_6 \longrightarrow 3N_2 + 5H_2 + 12CO + 2C$$

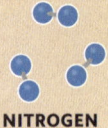
NITROGEN

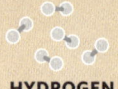
HYDROGEN

Products
The explosion of TNT forms three gases and carbon.

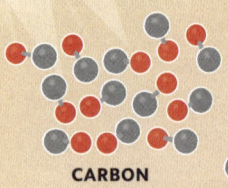
CARBON MONOXIDE

CARBON

68 | COMBUSTION AND EXPLOSIONS

DRIVING REACTIONS WITH ELECTRICITY

Electrolysis uses electricity to make substances react. The substances have to be either dissolved or melted, allowing charged particles (ions) to move. One simple example involves the formation of chlorine gas (Cl_2) and sodium metal (Na) when a positively charged electrode (anode) and a negatively charged electrode (cathode) are placed in molten sodium chloride (NaCl). The metal forms at the cathode, the nonmetal at the anode.

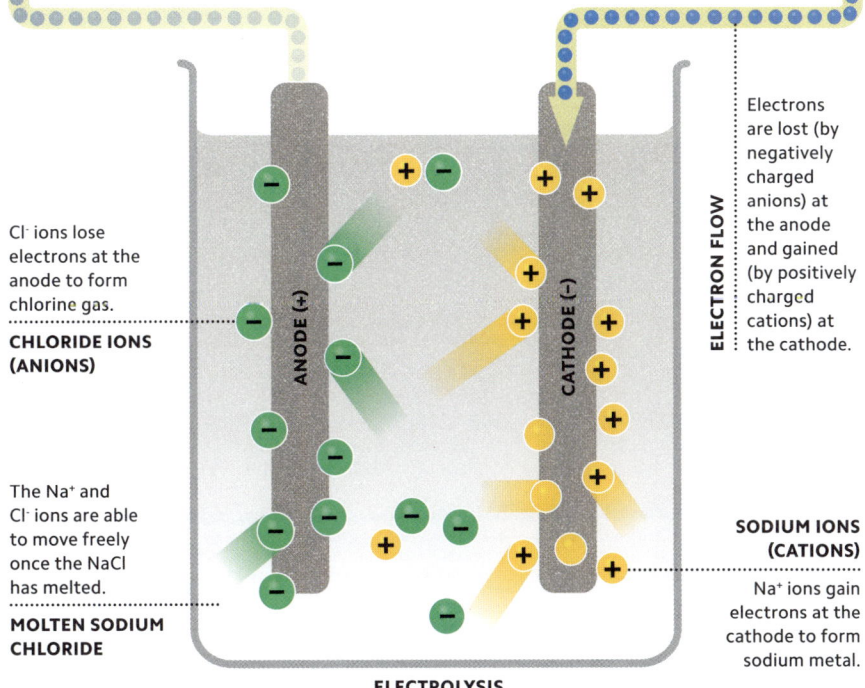

DIRECT CURRENT POWER SOURCE

CHLORIDE IONS (ANIONS)
Cl^- ions lose electrons at the anode to form chlorine gas.

MOLTEN SODIUM CHLORIDE
The Na^+ and Cl^- ions are able to move freely once the NaCl has melted.

SODIUM IONS (CATIONS)
Na^+ ions gain electrons at the cathode to form sodium metal.

Electrons are lost (by negatively charged anions) at the anode and gained (by positively charged cations) at the cathode.

ELECTRON FLOW

ANODE (+) CATHODE (−)

ELECTROLYSIS

THE SILENT MATCHMAKERS

Catalysts are substances that speed up chemical reactions without themselves being used up. They sometimes do this by holding one of the reactants in place, so that it can combine more readily with another substance. Enzymes (see p.105) are biological catalysts that speed up chemical reactions that need to happen in living things. In cars, catalytic converters help clean up exhaust fumes, converting toxic substances, such as carbon monoxide, into less dangerous substances.

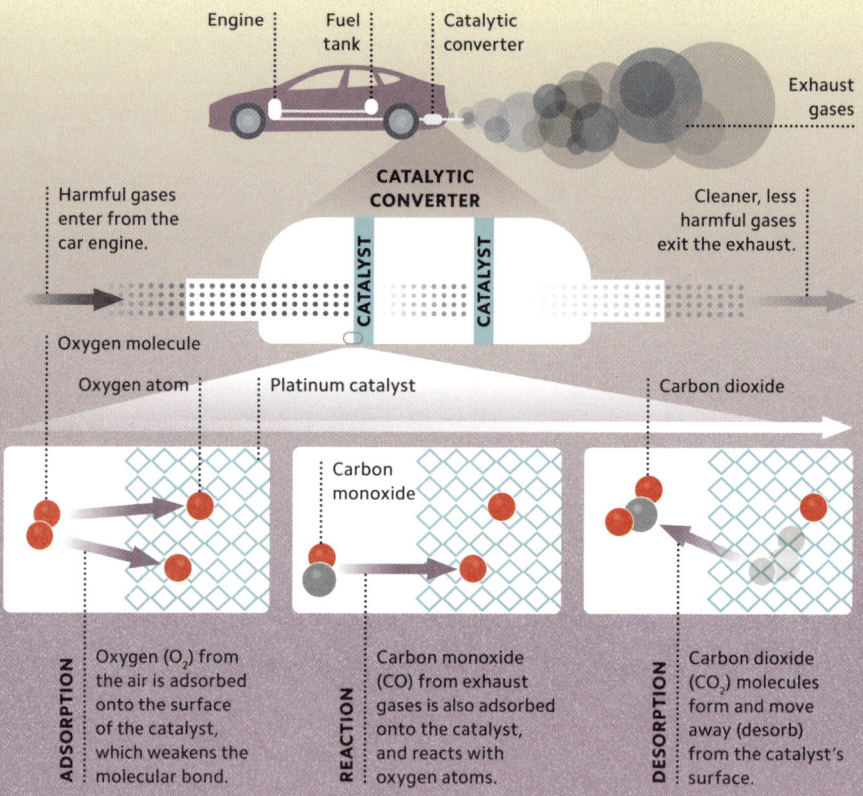

ADSORPTION — Oxygen (O_2) from the air is adsorbed onto the surface of the catalyst, which weakens the molecular bond.

REACTION — Carbon monoxide (CO) from exhaust gases is also adsorbed onto the catalyst, and reacts with oxygen atoms.

DESORPTION — Carbon dioxide (CO_2) molecules form and move away (desorb) from the catalyst's surface.

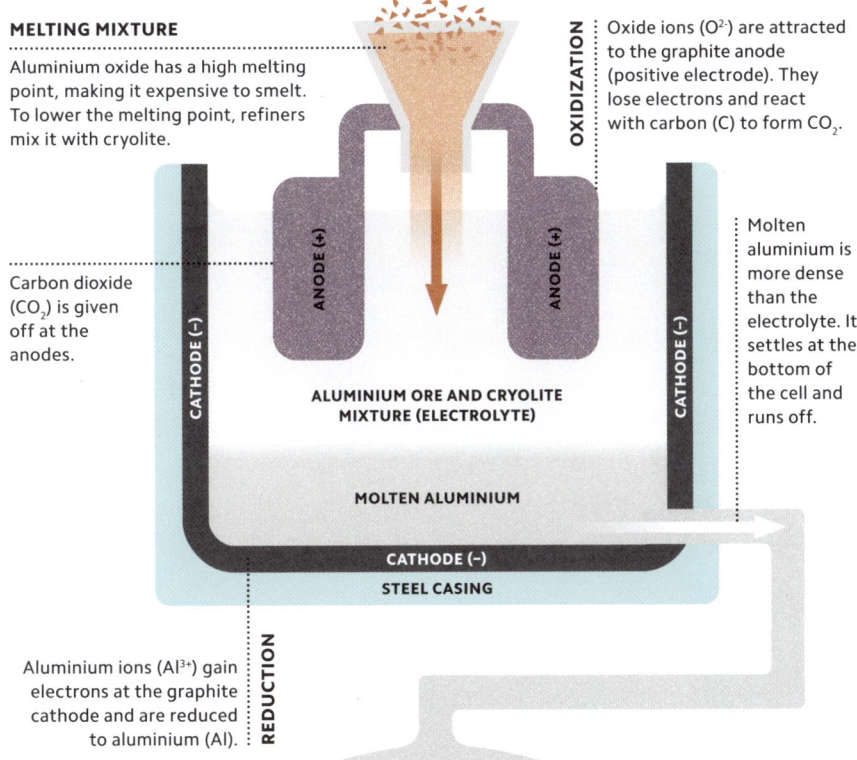

MELTING MIXTURE

Aluminium oxide has a high melting point, making it expensive to smelt. To lower the melting point, refiners mix it with cryolite.

OXIDIZATION

Oxide ions (O^{2-}) are attracted to the graphite anode (positive electrode). They lose electrons and react with carbon (C) to form CO_2.

Carbon dioxide (CO_2) is given off at the anodes.

Molten aluminium is more dense than the electrolyte. It settles at the bottom of the cell and runs off.

REDUCTION

Aluminium ions (Al^{3+}) gain electrons at the graphite cathode and are reduced to aluminium (Al).

PURIFYING EARTH'S TREASURE

Iron can be extracted from iron oxides through traditional chemical reduction (see p.67), but this does not work for more reactive metals, which must be extracted through electrolysis (see p.69). Aluminium is a very important metal, but extracting it from its ores requires a specific process. Aluminium ore is dissolved in molten cryolite to lower its melting point. The process releases greenhouse gases such as carbon dioxide (CO_2) and drives air pollution and climate change.

METAL REFINING

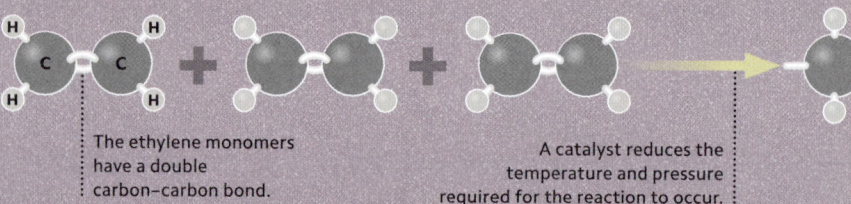

The ethylene monomers have a double carbon–carbon bond.

A catalyst reduces the temperature and pressure required for the reaction to occur.

BUILDING SUPERMOLECULES

Polymers are large molecules made from many smaller ones called monomers. In addition polymerization, heating the monomers causes their carbon–carbon double bonds to break, freeing them to make single bonds with other monomers and form a long chain. A catalyst speeds up the reaction and improves the quality of the polymer produced. The most well-known addition polymer is polyethylene (also called polythene), which is used to make many everyday items such as plastic bottles and bags.

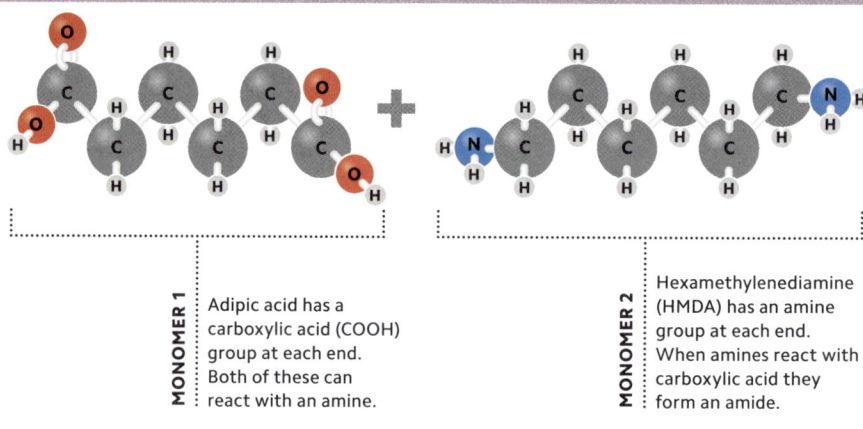

MONOMER 1 Adipic acid has a carboxylic acid (COOH) group at each end. Both of these can react with an amine.

MONOMER 2 Hexamethylenediamine (HMDA) has an amine group at each end. When amines react with carboxylic acid they form an amide.

Something extra
In the production of nylon, a water molecule (H_2O) is formed each time a pair of monomers join to make a section of polymer.

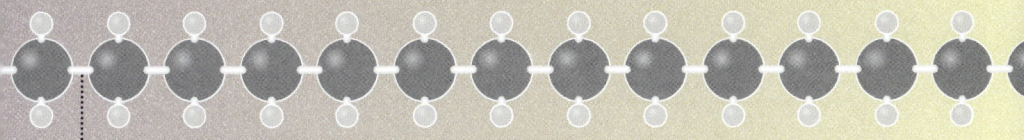

Polyethylene (the polymer) contains a long chain of carbon–carbon single bonds.

Using all molecules
Many small molecules (monomers) combine to make a larger polymer molecule, without forming any other products.

> "I am inclined to think that the development of polymerization is, perhaps, the biggest thing chemistry has done."
> Alexander Todd

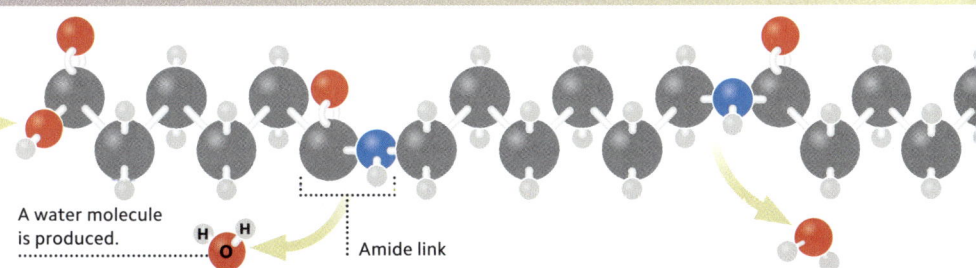

A water molecule is produced.

Amide link

MAKING MOLECULAR GIANTS

Unlike addition polymerization, condensation polymerization produces two products and involves pairs of different monomers. The two monomers need to have different functional groups, one at each end, responsible for the characteristic reactions of the compound. Examples of functional groups are acids and amines. These two groups react to form an amide link (a type of chemical bond).

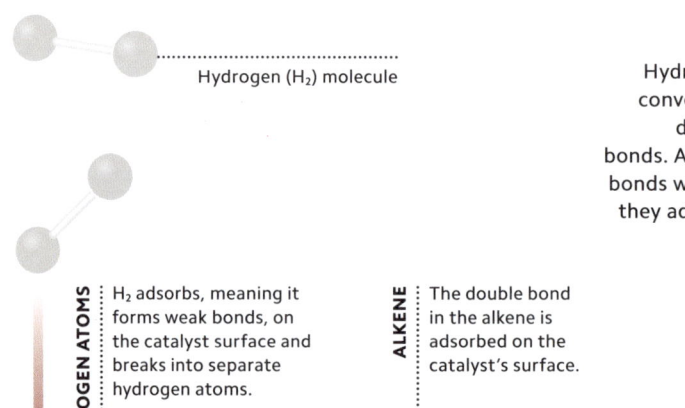

Hydrogen (H₂) molecule

Adding hydrogen
Hydrogenation is a way of converting carbon–carbon double bonds to single bonds. A catalyst weakens the bonds within the reactants as they adsorb onto its surface.

HYDROGEN ATOMS — H₂ adsorbs, meaning it forms weak bonds, on the catalyst surface and breaks into separate hydrogen atoms.

ALKENE — The double bond in the alkene is adsorbed on the catalyst's surface.

COLLISION — The alkene is on a collision course with the atoms of hydrogen.

NICKEL CATALYST SURFACE

REACTANT 1

IN CONTINUOUS MOTION

REACTIONS

Mixing two substances in a flask and waiting for them to react is called "batch chemistry". This makes a fixed amount of product before needing to restart. To produce substances on an industrial scale, "flow chemistry" is more useful. In a flow reaction, reactants continuously flow into a reactor while products are removed. Flow chemistry is typically safer, in part because smaller volumes are reacting at any one time and it is possible to monitor and adjust conditions such as pressure. The Haber process (see p.126) is an example of an industrial reaction that works this way.

REACTANT 2

74 | FLOW CHEMISTRY

HEAPS OF HYDROGEN

Hydrogenation is the reaction between hydrogen (H₂) and another compound, such as an alkene (see p.98). Usually, a carbon–carbon double bond becomes a single bond as the molecule gains two hydrogen atoms. This forms a "saturated" compound, where each carbon is bonded to the maximum number of atoms it can hold. Hydrogenation was once used in the food industry to make plant fats more solid at room temperature.

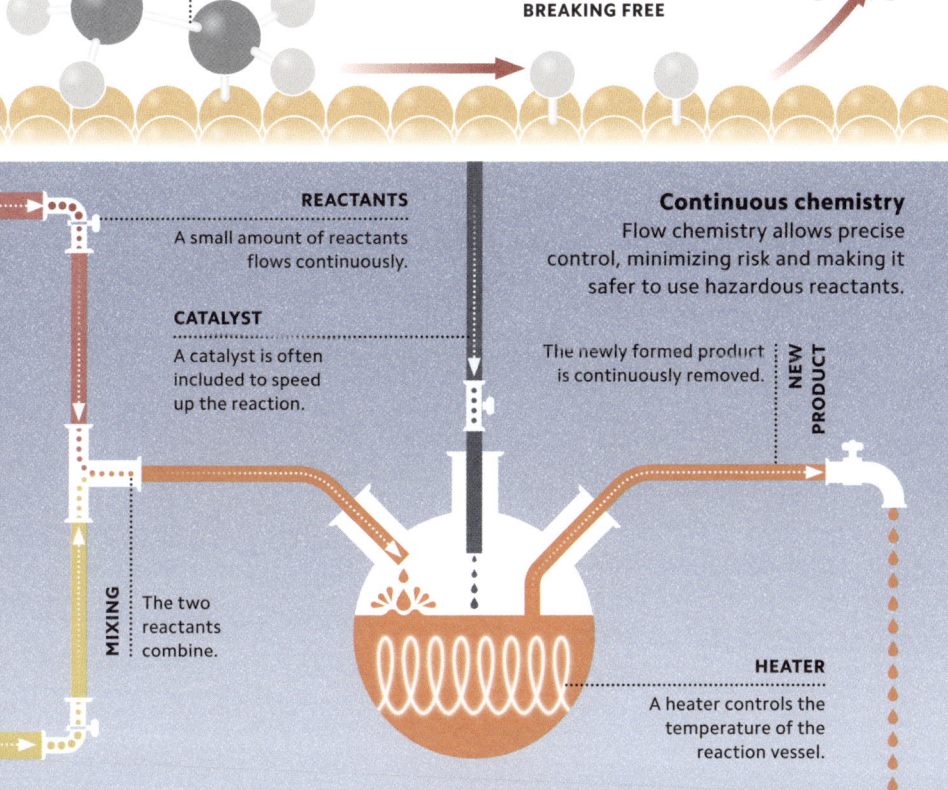

BONDS CHANGE
A collision causes a hydrogen atom to bond to a carbon atom in the alkane.

A second hydrogen reacts and produces a saturated hydrocarbon molecule that leaves the metal's surface.
BREAKING FREE

REACTANTS
A small amount of reactants flows continuously.

CATALYST
A catalyst is often included to speed up the reaction.

MIXING
The two reactants combine.

Continuous chemistry
Flow chemistry allows precise control, minimizing risk and making it safer to use hazardous reactants.

The newly formed product is continuously removed.
NEW PRODUCT

HEATER
A heater controls the temperature of the reaction vessel.

SEPARATING SUBSTANCES WITH HEAT

Different substances have different boiling points, and this can be used to separate them when they are in a mixture. Crude oil is a mixture of hydrocarbon molecules made from hydrogen and carbon of different lengths – the shorter the chain, the lower the boiling point. A fractional distillation column heats crude oil. Substances with the shortest chains, such as petrol, rise to the top of the column and condense. Longer chains, with higher boiling points, leave the bottom of the column.

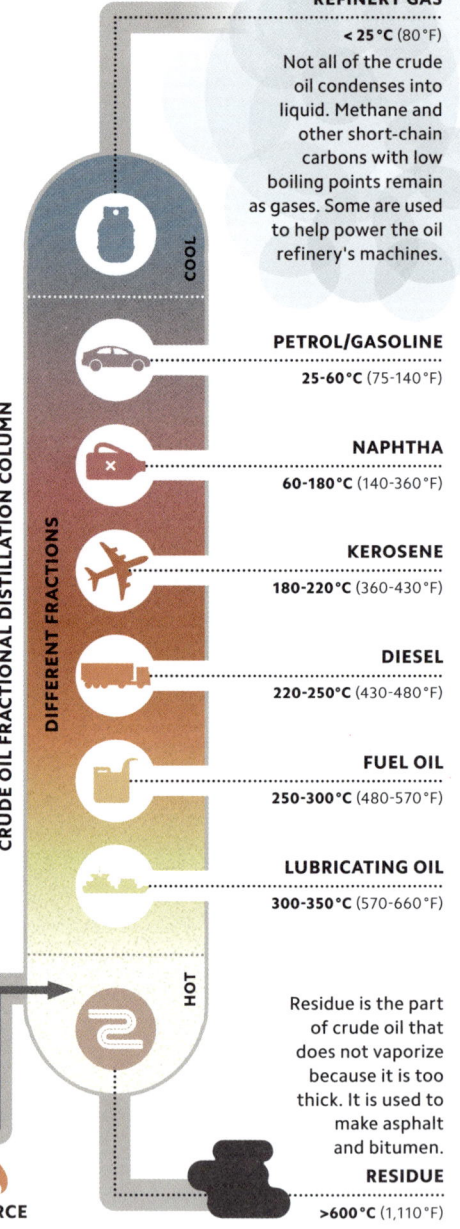

REFINERY GAS
< 25 °C (80 °F)
Not all of the crude oil condenses into liquid. Methane and other short-chain carbons with low boiling points remain as gases. Some are used to help power the oil refinery's machines.

PETROL/GASOLINE
25-60 °C (75-140 °F)

NAPHTHA
60-180 °C (140-360 °F)

KEROSENE
180-220 °C (360-430 °F)

DIESEL
220-250 °C (430-480 °F)

FUEL OIL
250-300 °C (480-570 °F)

LUBRICATING OIL
300-350 °C (570-660 °F)

Residue is the part of crude oil that does not vaporize because it is too thick. It is used to make asphalt and bitumen.
RESIDUE
>600 °C (1,110 °F)

CRUDE OIL   **HEAT SOURCE**

CRUDE OIL FRACTIONAL DISTILLATION COLUMN — DIFFERENT FRACTIONS — COOL — HOT

COLOURFUL CLOCKS

Most chemical reactions start and then stop. Some may be set up to react continuously, but a few fluctuate over time. A chemical oscillator is a mixture of reacting compounds in which the concentration of one or more of the substances goes up and down with time. Often, a product is also a reactant, so a cycle occurs, which may be visible as a colour change that comes and goes. Such reactions are useful for studying the rates of chemical reactions and how patterns such as fingerprints form.

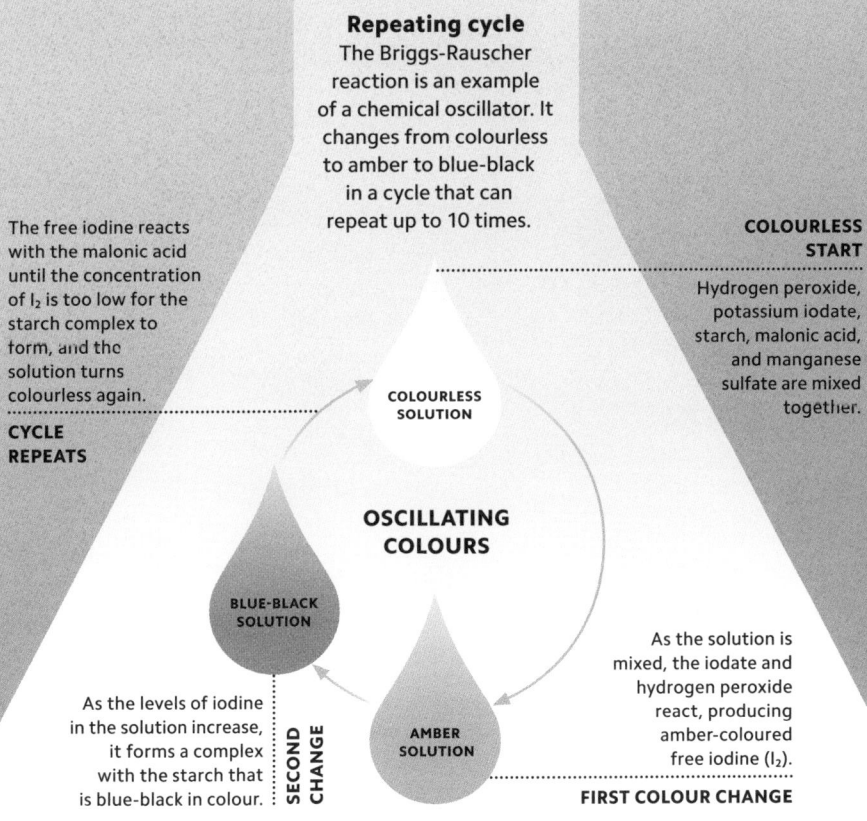

Repeating cycle
The Briggs-Rauscher reaction is an example of a chemical oscillator. It changes from colourless to amber to blue-black in a cycle that can repeat up to 10 times.

COLOURLESS START
Hydrogen peroxide, potassium iodate, starch, malonic acid, and manganese sulfate are mixed together.

OSCILLATING COLOURS

COLOURLESS SOLUTION

BLUE-BLACK SOLUTION

AMBER SOLUTION

FIRST COLOUR CHANGE
As the solution is mixed, the iodate and hydrogen peroxide react, producing amber-coloured free iodine (I_2).

SECOND CHANGE
As the levels of iodine in the solution increase, it forms a complex with the starch that is blue-black in colour.

CYCLE REPEATS
The free iodine reacts with the malonic acid until the concentration of I_2 is too low for the starch complex to form, and the solution turns colourless again.

CHEMICAL OSCILLATORS | 77

ANALY
CHEMI

To determine what a substance is, and how much of it is present in a mixture or compound, demands analytical insights of enormous ingenuity and flair. In centuries gone by, chemists developed methods of analysis using colour, pattern, and other observable properties, including even taste and smell. Modern-day analytical chemistry still relies on skilled observations and experimental enquiries, alongside complex instrumental methods including spectroscopy, crystallography, microscopy, and mass spectrometry. Such investigative tools enable environmental monitoring, manufacturing, and forensics, as well as the analysis of foods and medicines for safety and quality.

PULLING MIXTURES APART

Chromatography is a way of separating mixtures of soluble substances. A simple example involves putting small spots of ink onto absorbent paper, placing the bottom of the paper in a solvent, such as water, and waiting for the solvent to travel up the paper. The particles that make up the ink get caught in the moving solvent. Some particles move more quickly than others, depending on how strongly they are held by the paper, leading the mixture to separate. The resulting pattern is called a chromatogram.

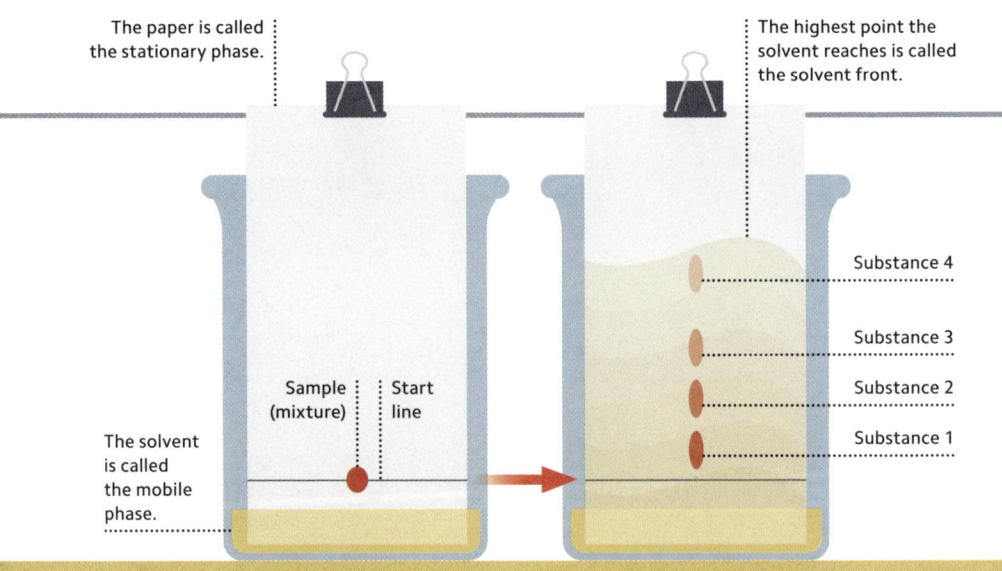

Sample
The sample, such as an ink, which contains a mixture of different substances, is dotted on a pencil line. The paper is suspended in a solvent.

Separation
As the solvent travels up the paper, it causes the substances in the mixture to separate because they hold on to the paper with different strengths.

Jumping electrons
The electron transitions are specific to different metals, resulting in characteristic colours for different metal ions.

BURNING BRIGHTLY

POTASSIUM

COPPER

SODIUM

LITHIUM

BARIUM

BARIUM IONS The electrons in barium ions (Ba^{2+}) release light energy with a wavelength of 554 nanometers (nm): a light apple green colour.

Some substances can change the colour of a flame due to the presence of metal ions. The electrons in metal ions jump into higher energy levels when heated, then release energy as light at distinct wavelengths and hence colours, when they fall back down. These colours are consistent, so this can be used to identify a particular metal in a substance. For example, copper ions (Cu^{2+}) produce a striking blue-green flame, while potassium ions (K^+) produce a lilac flame.

DETECTIVE WORK

Some chemical reactions produce changes that are easy to see or hear, and as a result they can help chemists work out whether particular substances are present or absent. These tests are qualitative, meaning they are based on observations rather than precise measurements. With these tests chemists can, for example, confirm the presence of carbon dioxide, identify oxygen or hydrogen, or determine pH with the use of indicators.

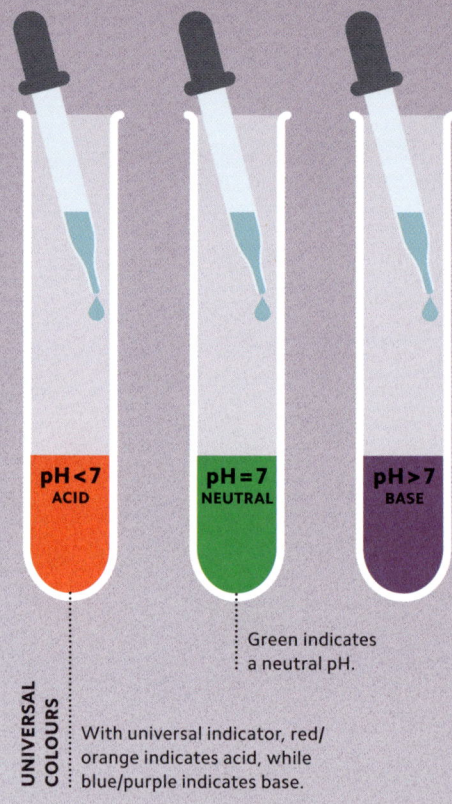

UNIVERSAL COLOURS

With universal indicator, red/orange indicates acid, while blue/purple indicates base.

Green indicates a neutral pH.

> Litmus indicator dyes are extracted from various species of lichen.

Measuring pH
Indicators are dyes that change colour at different pHs (see p.47). Universal indicator is actually a mixture of several different indicators, allowing pH levels from 0 to 14 on the pH scale (see p.46) to be visually identified.

82 | CHEMICAL ANALYSIS TESTS

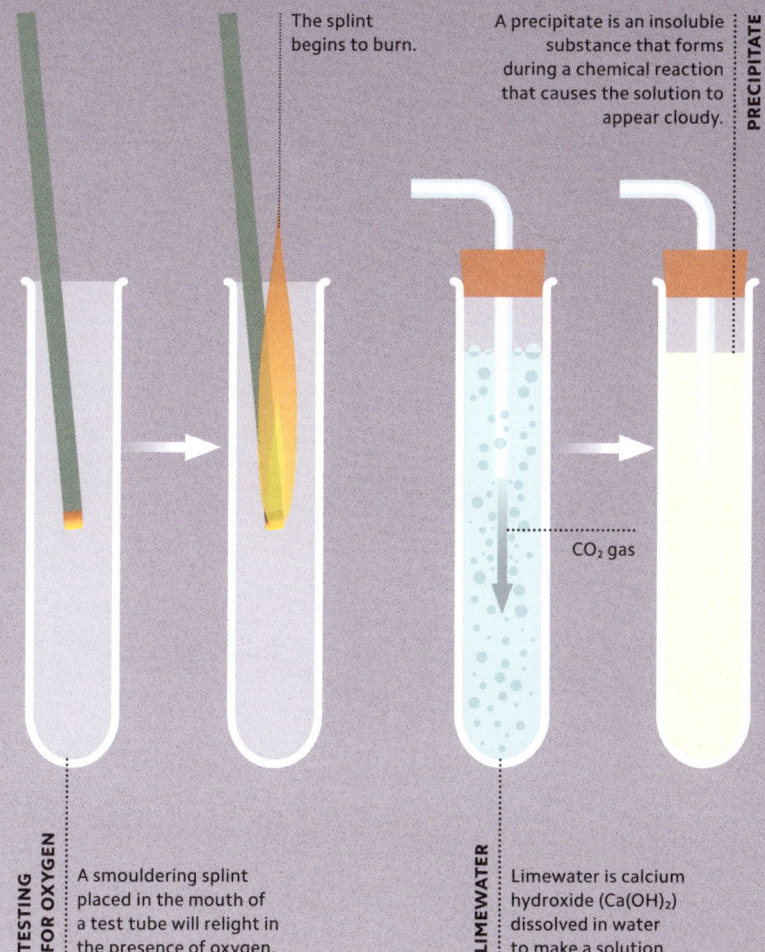

The splint begins to burn.

A precipitate is an insoluble substance that forms during a chemical reaction that causes the solution to appear cloudy.

PRECIPITATE

TESTING FOR OXYGEN: A smouldering splint placed in the mouth of a test tube will relight in the presence of oxygen.

CO_2 gas

LIMEWATER: Limewater is calcium hydroxide ($Ca(OH)_2$) dissolved in water to make a solution.

Identifying gases
It is useful to be able to identify gases produced during chemical reactions. Oxygen gas (O_2) causes a glowing splint to relight. Hydrogen gas (H_2) produces a squeaky pop sound. Chlorine gas (Cl_2) bleaches damp blue litmus paper.

Limewater test
When carbon dioxide (CO_2) bubbles through limewater, a white precipitate of calcium carbonate ($CaCO_3$) forms and the limewater turns cloudy or "milky".

CHEMICAL ANALYSIS TESTS

HOT GAS

A voltage passes through a glass tube of low-pressure mercury gas, causing it to glow.

MERCURY GAS

SLIT

Light emitted by the heated gas passes though a slit, producing a narrow beam.

FINGERPRINTING ELEMENTS

Emission spectroscopy makes use of the fact that some elements emit coloured light when they are hot (see p.81). Instead of relying on eyesight to identify colours, a machine called a spectroscope measures the wavelengths and intensity of the light and produces a "line spectrum", which can be compared to reference spectra to identify an element.

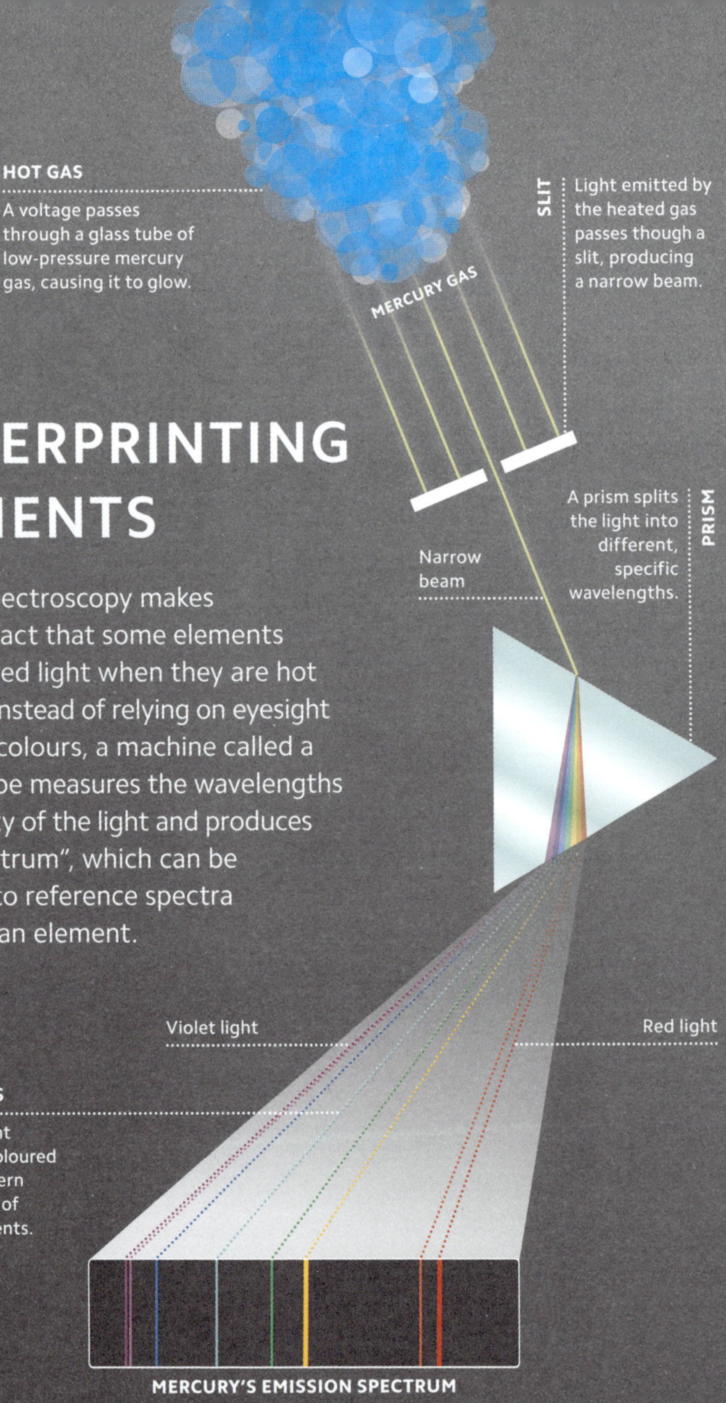

Narrow beam

PRISM

A prism splits the light into different, specific wavelengths.

Violet light

Red light

SEPARATED WAVELENGTHS

The emitted light forms narrow, coloured bands. The pattern is characteristic of particular elements.

MERCURY'S EMISSION SPECTRUM

84 | EMISSION SPECTROSCOPY

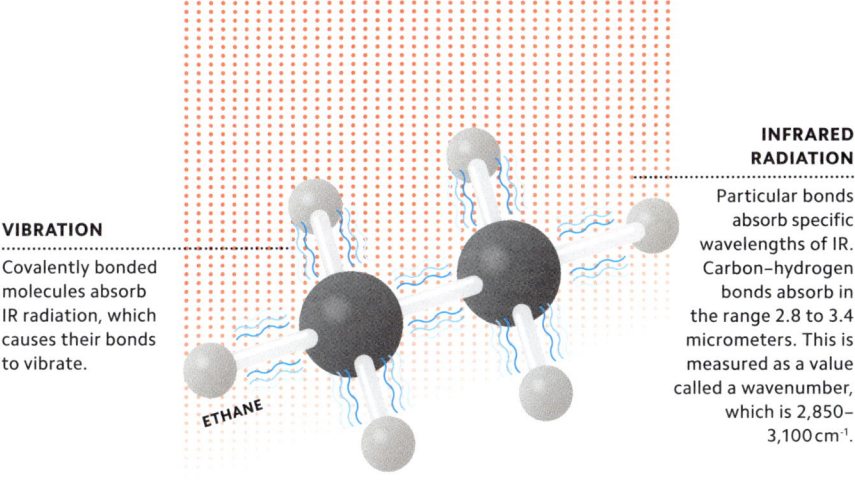

VIBRATION

Covalently bonded molecules absorb IR radiation, which causes their bonds to vibrate.

ETHANE

INFRARED RADIATION

Particular bonds absorb specific wavelengths of IR. Carbon–hydrogen bonds absorb in the range 2.8 to 3.4 micrometers. This is measured as a value called a wavenumber, which is 2,850–3,100 cm^{-1}.

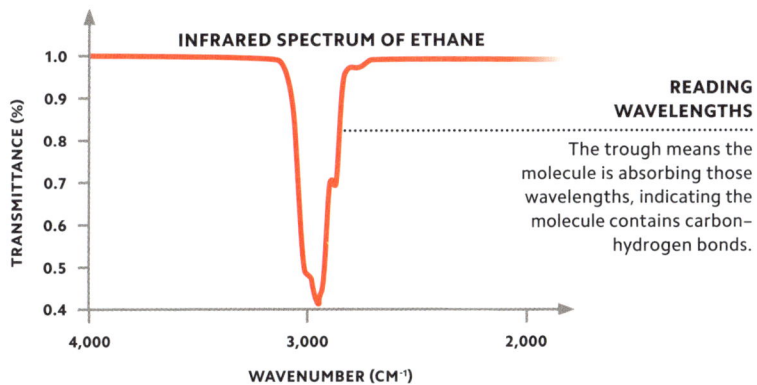

INFRARED SPECTRUM OF ETHANE

READING WAVELENGTHS

The trough means the molecule is absorbing those wavelengths, indicating the molecule contains carbon–hydrogen bonds.

MOLECULAR MOTION

Infrared (IR) radiation is the part of the electromagnetic spectrum that sits between visible red light and microwaves. Molecular bonds vibrate as they absorb certain IR wavelengths. A sample is put into a spectrometer, which beams IR radiation through it, and a spectrum shows exactly which wavelengths were absorbed, and which passed through the sample. This can indicate what types of bonds are present, and helps chemists identify substances such as organic molecules.

WEIGHING
THE INVISIBLE

A mass spectrometer is a piece of equipment that measures the mass of molecules or atoms. Mass spectrometers knock one or more electrons off an atom or molecule to produce ions, which a magnetic field then redirects, or deflects. The amount of deflection – measured by the detector – depends on the mass and how many electrons are knocked off. Organic molecules also break up in distinctive ways, which makes it possible to identify specific substances.

Measuring masses
A mass spectrometer produces positively charged ions in a vacuum, then accelerates, and deflects them. It measures the time it takes the ions to reach the detector.

ACCELERATING PLATE
Positively charged ions are accelerated due to their attraction to electrically charged plates.

ELECTRON TRAP
A positively charged plate attracts electrons from the coil.

Sample

Sample injection port

DETECTOR
The ions meeting the detector result in a tiny electrical signal that scientists read on a computer.

ELECTROMAGNETS
The magnetic field deflects the ions: the lighter they are, the more they deflect.

HEATED METAL COIL
A heated metal coil gives off electrons that bombard the sample particles, turning them into ions.

ALIGNED
In a magnetic field, the nuclei line up, either with or against the field.

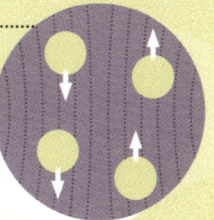

RADIO WAVE PULSE
Radio waves of a specific frequency "flip" the nuclei from one orientation to the other.

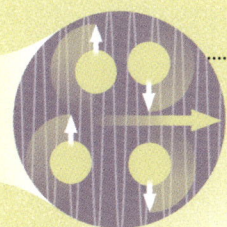

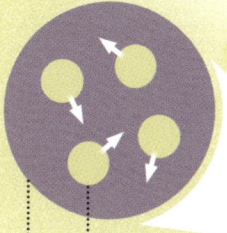

RANDOM

Protons are randomly aligned.

Hydrogen nuclei (protons) behave as though they spin on their axis, and, as a result behave like tiny magnets.

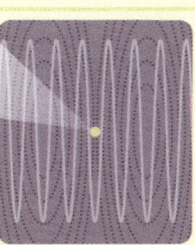

PROTON NUCLEAR MAGNETIC RESONANCE (NMR)

As nuclei "relax" back to their original orientation they release their own weak radio signal. Measuring this shows how atoms are connected.

NUCLEI RELAX

NUCLEI IN MAGNETIC FIELDS

Nuclear magnetic resonance (NMR) is a particularly powerful technique for identifying organic carbon-based substances. It works because some atomic nuclei, notably hydrogen-1, interact with powerful magnetic fields and radio waves. The technique provides detailed information – often enabling chemists to work out the structures of unknown substances. Magnetic resonance imaging (MRI) scanners use NMR to create detailed images of the body.

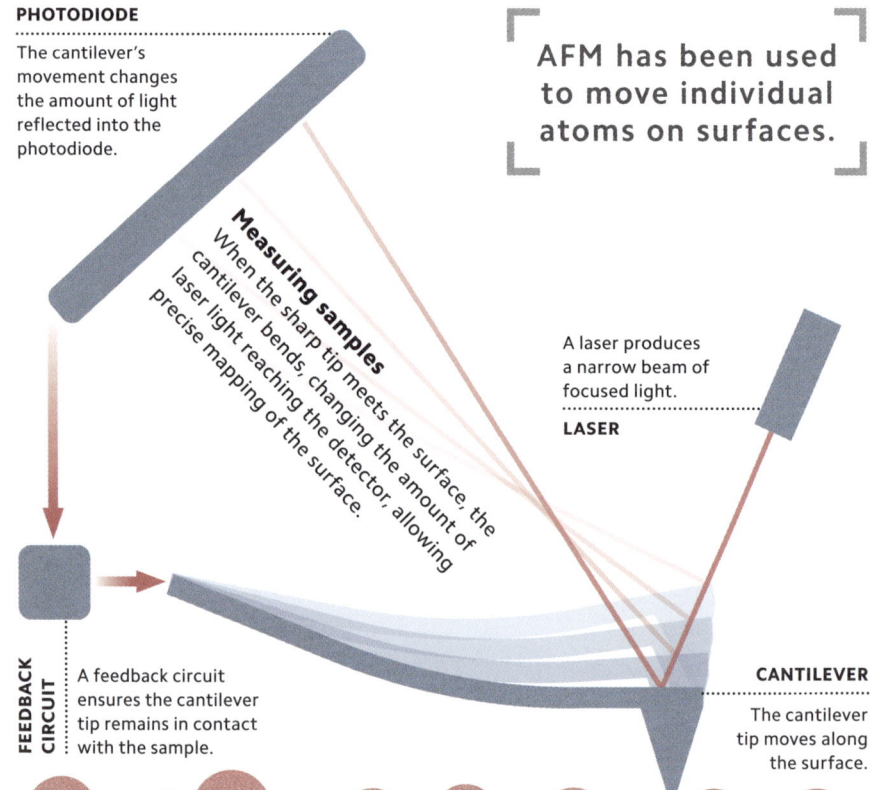

PHOTODIODE
The cantilever's movement changes the amount of light reflected into the photodiode.

AFM has been used to move individual atoms on surfaces.

Measuring samples
When the sharp tip meets the surface, the cantilever bends, changing the amount of laser light reaching the detector, allowing precise mapping of the surface.

A laser produces a narrow beam of focused light.
LASER

FEEDBACK CIRCUIT
A feedback circuit ensures the cantilever tip remains in contact with the sample.

CANTILEVER
The cantilever tip moves along the surface.

MAPPING SURFACES

Atomic force microscopy (AFM) uses a tiny probe to "feel" surfaces on the atomic level. This allows scientists to make pictures of atoms and molecules – and even molecular bonds. It can also identify atoms: for example, detecting the difference between silicon, tin, and lead on the surface of an alloy. The downsides are that only tiny samples, which have been carefully prepared, can be scanned, and getting a good-quality image can be time-consuming.

FREEZING VISION

In cryogenic electron microscopy (cryo-EM) a beam of electrons is transmitted through a very cold, ultrathin biological sample. An image forms due to the interaction between electrons that the atoms in the sample scatter and electrons that pass straight between the atoms. Using cryogenic temperatures (below −150°C/−238°F) reduces the damage the electron beam might otherwise do to the sample. It is an important technique in biochemistry, where it has been used to work out the 3D structures of complex biological molecules.

Observing molecules
Unlike X-ray crystallography (see p.90), this method does not require 3D crystals, which means complicated biomolecules can be observed in their more "natural" positions.

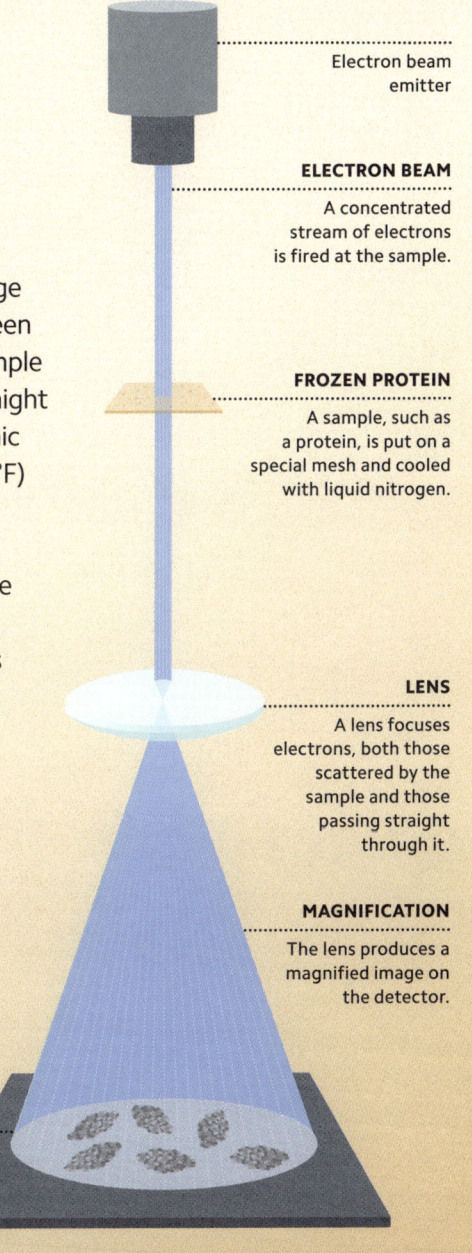

Electron beam emitter

ELECTRON BEAM
A concentrated stream of electrons is fired at the sample.

FROZEN PROTEIN
A sample, such as a protein, is put on a special mesh and cooled with liquid nitrogen.

LENS
A lens focuses electrons, both those scattered by the sample and those passing straight through it.

MAGNIFICATION
The lens produces a magnified image on the detector.

ELECTRON DETECTOR
A computer interprets the traces of hundreds of 2D images created to form a 3D model of the molecule.

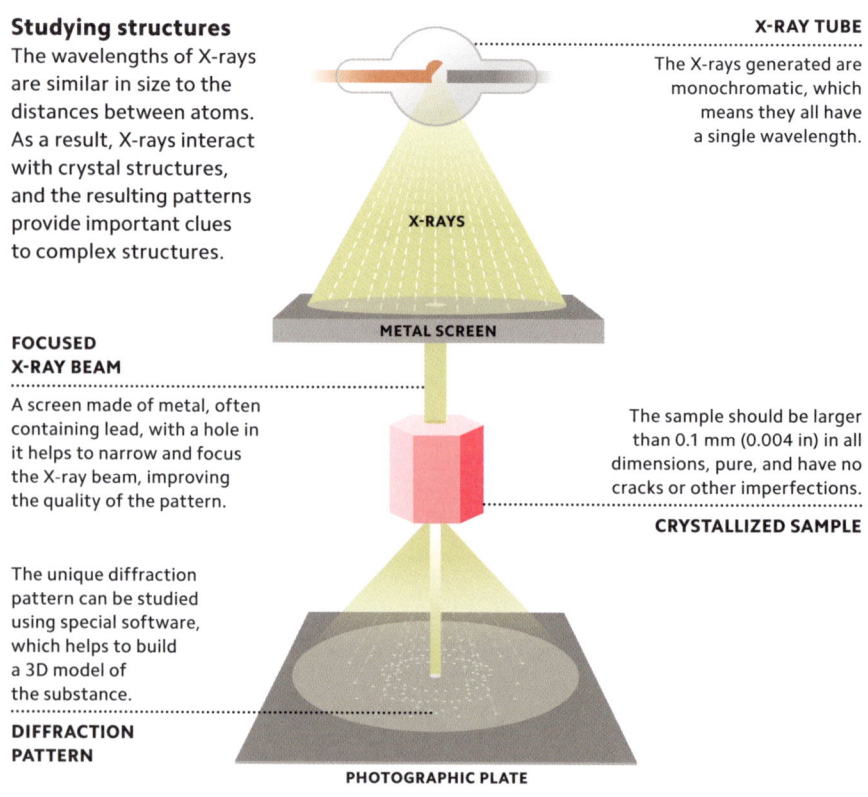

Studying structures
The wavelengths of X-rays are similar in size to the distances between atoms. As a result, X-rays interact with crystal structures, and the resulting patterns provide important clues to complex structures.

FOCUSED X-RAY BEAM
A screen made of metal, often containing lead, with a hole in it helps to narrow and focus the X-ray beam, improving the quality of the pattern.

The unique diffraction pattern can be studied using special software, which helps to build a 3D model of the substance.
DIFFRACTION PATTERN

X-RAY TUBE
The X-rays generated are monochromatic, which means they all have a single wavelength.

The sample should be larger than 0.1 mm (0.004 in) in all dimensions, pure, and have no cracks or other imperfections.
CRYSTALLIZED SAMPLE

EXAMINING CRYSTALLINE SUBSTANCES

X-ray crystallography is a technique in which X-rays are fired at a crystalline form of a substance. The X-rays scatter in specific directions and, by measuring the angles and intensities in the resulting pattern, the 3D structure of the substance can be determined. This technique was first used to work out the sizes of atoms and lengths of chemical bonds, and later to reveal the structure of many biological molecules, including drugs, vitamins, and, famously, DNA.

GAS SLEUTHING

Gas chromatography is a special type of chromatography (see p.80) used to separate compounds that vaporize easily at room temperature. A sample mixed with an inert gas, such as helium, forms the mobile phase. This mixture is passed along the stationary phase – a long, narrow tube that is lined with an immobile material, often silica. The interaction of the mobile and stationary phases causes the substances in the mixture to separate. An electron capture detector accurately identifies substances and is especially sensitive to electronegative groups, such as halogen atoms found in pesticides.

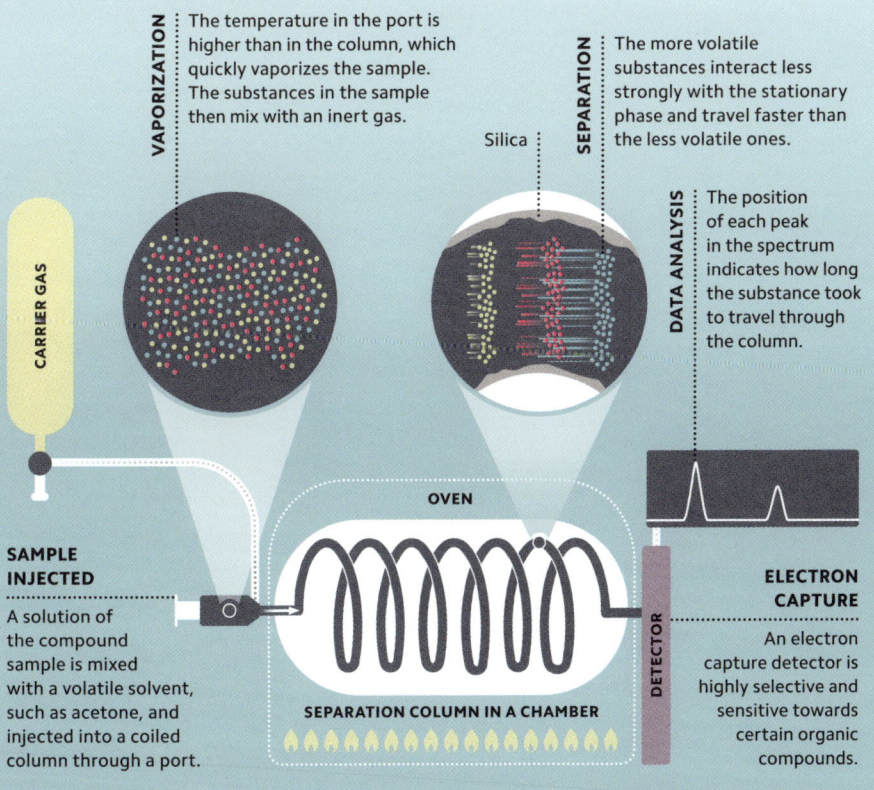

VAPORIZATION
The temperature in the port is higher than in the column, which quickly vaporizes the sample. The substances in the sample then mix with an inert gas.

SEPARATION
The more volatile substances interact less strongly with the stationary phase and travel faster than the less volatile ones.

Silica

DATA ANALYSIS
The position of each peak in the spectrum indicates how long the substance took to travel through the column.

CARRIER GAS

SAMPLE INJECTED
A solution of the compound sample is mixed with a volatile solvent, such as acetone, and injected into a coiled column through a port.

OVEN

SEPARATION COLUMN IN A CHAMBER

DETECTOR

ELECTRON CAPTURE
An electron capture detector is highly selective and sensitive towards certain organic compounds.

GAS CHROMATOGRAPHY | 91

BIOOR
CHEMI

GANIC
STRY

To understand biological systems, you also need chemistry. Bioorganic chemists apply chemical methods to study biological processes. Bioorganic chemistry also explores what life is and how it arose. As a continuously evolving discipline, it uses insights from both biology and chemistry to shine light into new scientific corners. With the focus on living systems, bioorganic chemists examine organic molecules such as proteins, enzymes, and nucleic acids to understand their reactions and interactions. The resulting insights help in medicine and manufacturing, with ongoing research driving new advancements and applications, particularly in drug development and biotechnology.

THE SMART CHEMISTRY REVOLUTION

Researchers in all areas of chemistry are developing incredible artificial intelligence (AI) tools to fast-track new discoveries. These tools are already helping cut the carbon footprint produced by massive supercomputers doing complicated chemistry calculations. AI can help do these calculations more simply, using less energy. Using AI, scientists have also solved a decades-old problem in bioorganic chemistry – how to predict the complex structure of a protein from the amino acid sequence encoded in DNA.

> The developers of AI programs called RoseTTAFold and AlphaFold2 won the 2024 Nobel Prize in Chemistry.

LOTS OF LINKAGES

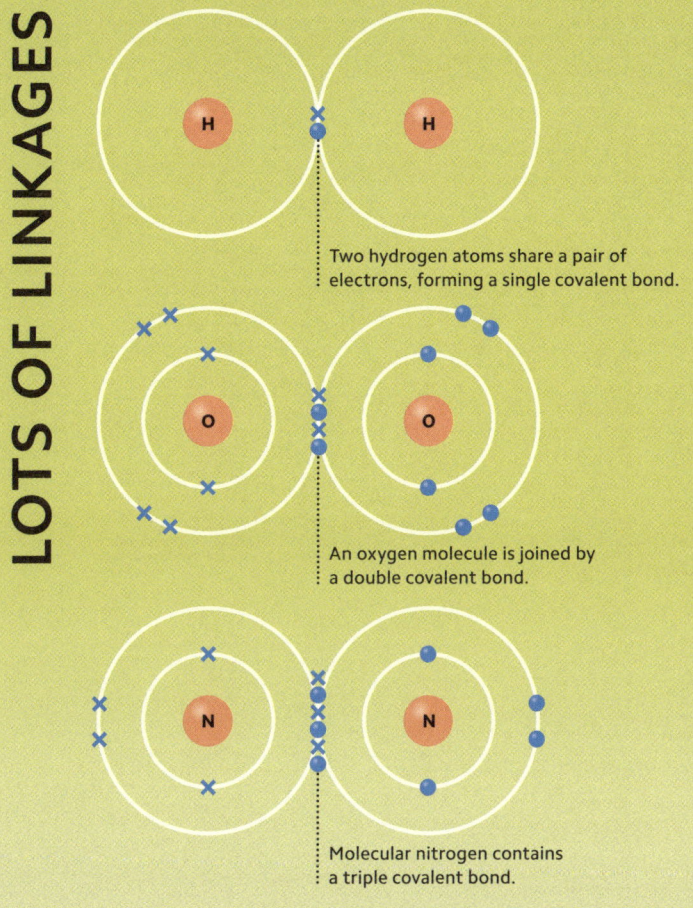

Two hydrogen atoms share a pair of electrons, forming a single covalent bond.

An oxygen molecule is joined by a double covalent bond.

Molecular nitrogen contains a triple covalent bond.

Covalent bonds (see pp.36–37) hold together atoms in almost all biological molecules. But the type of bond and the atoms involved have a huge impact on the properties of the resulting compound. Single bonds, in which each atom contributes one electron, are the most common but also the weakest. Double bonds use twice as many electrons, sharing four electrons between the two atoms, and often link two carbon atoms (as in alkenes) or carbon and oxygen. Triple bonds are the strongest and rarest as each atom gives three electrons.

SINGLE, DOUBLE, AND TRIPLE BONDS | 95

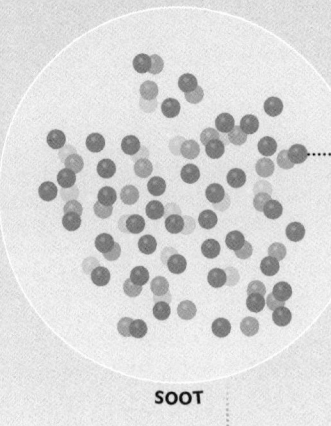

SOOT

The carbon atoms are arranged randomly, giving this allotrope a disorganized structure.

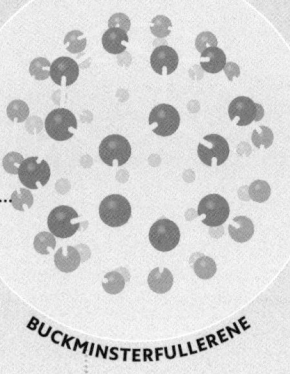

BUCKMINSTERFULLERENE

This spherical arrangement of 60 carbon atoms is formed of fused five- and six-sided rings, like a football.

VARYING FORMS

Sometimes the atoms that make up a particular element can bond together in different ways, creating different physical forms of the element, known as allotropes. Carbon is the most famous example. Both diamond and graphite are forms of pure carbon, but their atomic structure gives them extremely different properties. In diamond, every carbon atom is covalently bonded (see pp.36–37) to four other carbon atoms, creating a rigid 3D structure. By contrast, in graphite, each carbon atom bonds only to three other carbon atoms, forming flat sheets of hexagons, which slide across each other, making graphite soft.

CARBON

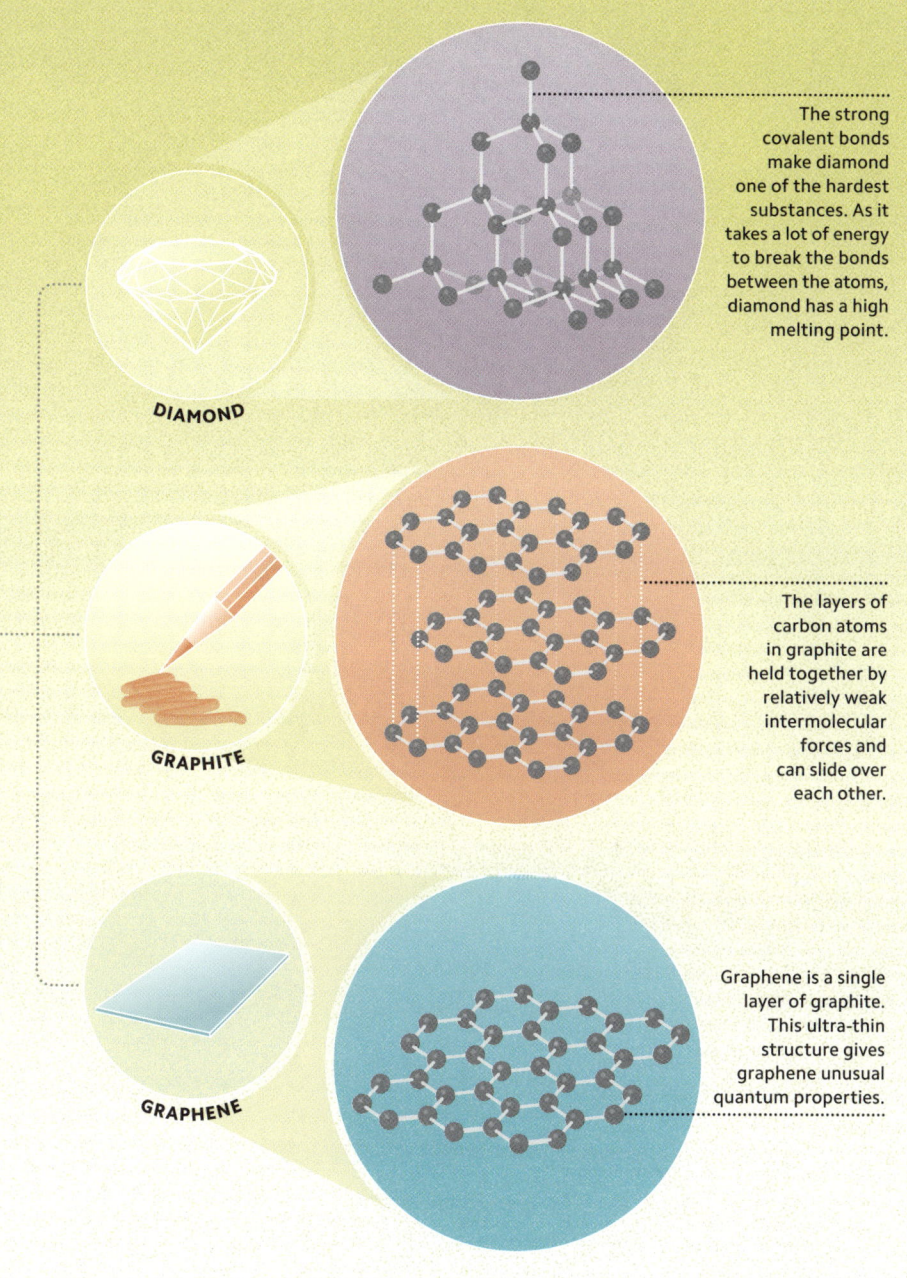

DIAMOND — The strong covalent bonds make diamond one of the hardest substances. As it takes a lot of energy to break the bonds between the atoms, diamond has a high melting point.

GRAPHITE — The layers of carbon atoms in graphite are held together by relatively weak intermolecular forces and can slide over each other.

GRAPHENE — Graphene is a single layer of graphite. This ultra-thin structure gives graphene unusual quantum properties.

ALLOTROPES

THE CARBON ZOO

Hydrocarbons, composed solely of hydrogen and carbon atoms, are a group of flammable compounds. The simplest members, alkanes, have single bonds between the carbon atoms, and multiple bonds to hydrogen atoms. They range from one-carbon methane (CH_4) to huge structures such as hexacontane ($C_{60}H_{122}$). Alkenes include at least one double bond within their structure, making them more reactive than alkanes. The final group, alkynes, contain a carbon–carbon triple bond.

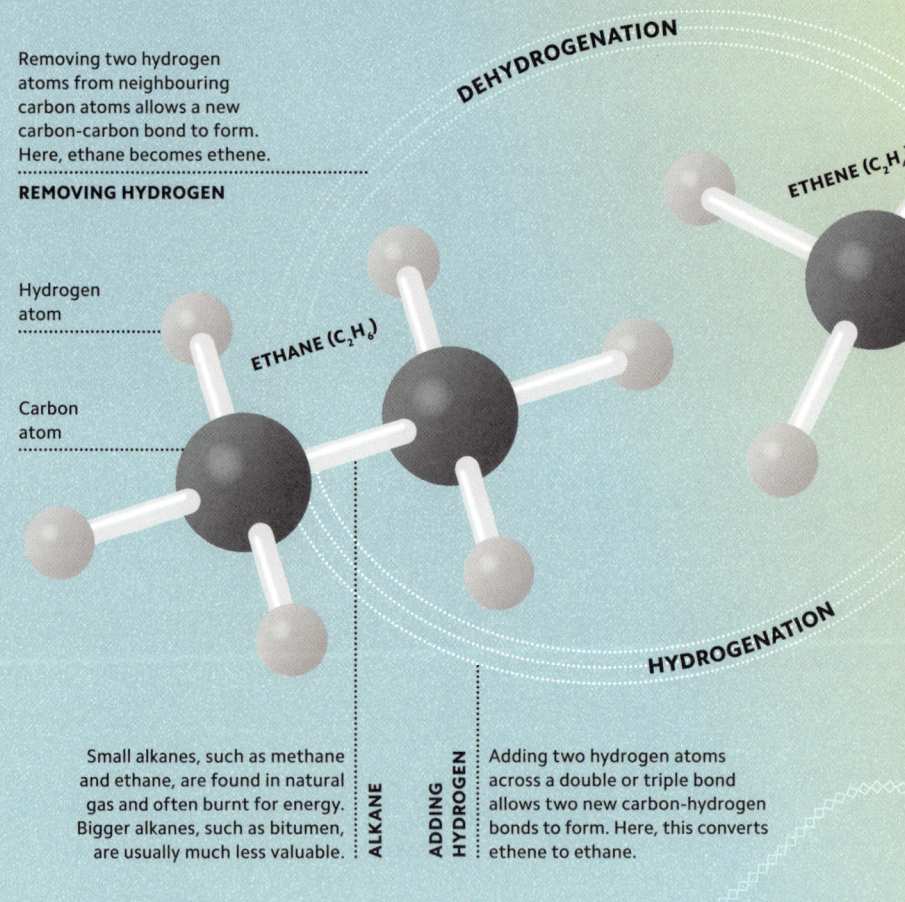

REMOVING HYDROGEN
Removing two hydrogen atoms from neighbouring carbon atoms allows a new carbon-carbon bond to form. Here, ethane becomes ethene.

Hydrogen atom

Carbon atom

ALKANE
Small alkanes, such as methane and ethane, are found in natural gas and often burnt for energy. Bigger alkanes, such as bitumen, are usually much less valuable.

ADDING HYDROGEN
Adding two hydrogen atoms across a double or triple bond allows two new carbon-hydrogen bonds to form. Here, this converts ethene to ethane.

98 | ALKANES, ALKENES, AND ALKYNES

> The carbon–hydrogen bonds of alkanes, alkenes, and alkynes store a lot of energy.

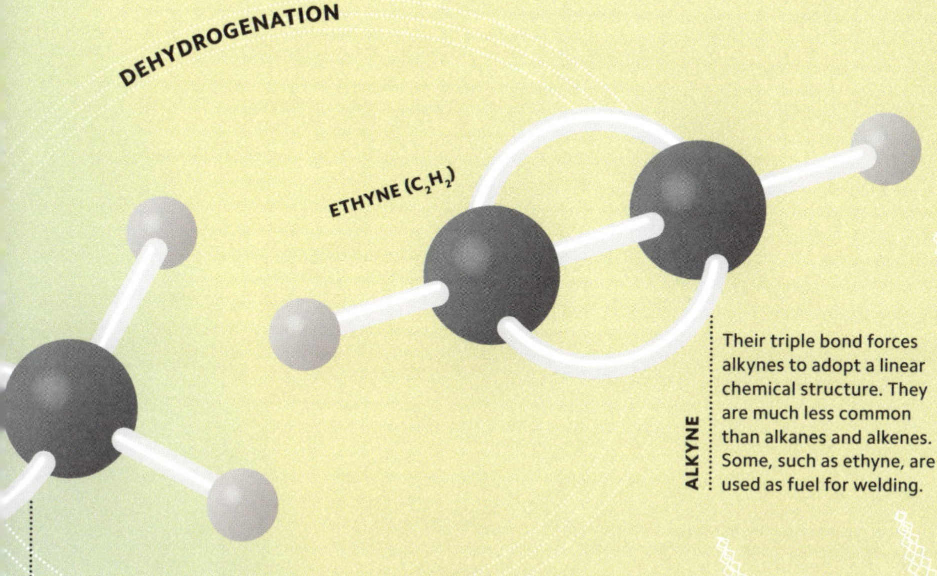

DEHYDROGENATION

ETHYNE (C_2H_2)

ALKYNE
Their triple bond forces alkynes to adopt a linear chemical structure. They are much less common than alkanes and alkenes. Some, such as ethyne, are used as fuel for welding.

HYDROGENATION

ALKENE
Alkenes are manufactured from crude oil through a process called cracking (see p.100) that breaks long-chain molecules into smaller, more useful pieces.

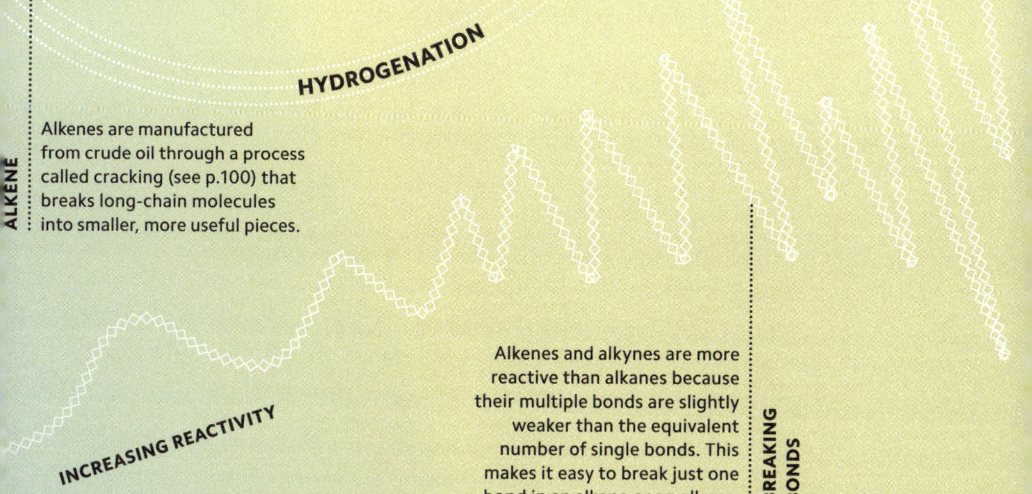

INCREASING REACTIVITY

BREAKING BONDS
Alkenes and alkynes are more reactive than alkanes because their multiple bonds are slightly weaker than the equivalent number of single bonds. This makes it easy to break just one bond in an alkene or an alkyne.

ALKANES, ALKENES, AND ALKYNES | 99

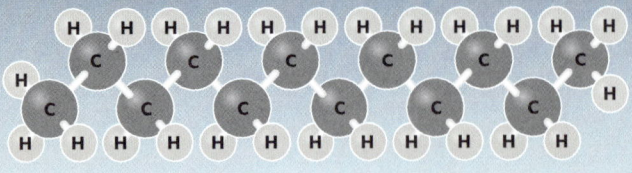

DODECANE $C_{12}H_{26}$ (ALKANE)

ZEOLITE CATALYST

The zeolite catalyst has a porous structure, creating plenty of sites for the hot hydrocarbon to bind to the surface and undergo the breakdown reaction.

BREAKING BONDS

Fractional distillation (see p.76) separates the different chemical components of crude oil by their boiling point. However, this process tends to produce more long-chain alkanes than useful small-chain hydrocarbons. A second process, called cracking, breaks down some of the long-chain alkanes into smaller hydrocarbon components. Catalytic cracking uses an aluminosilicate catalyst, usually in a form called a zeolite, to break the molecules apart at around 550°C (1,020°F). While steam cracking does not need a catalyst, it requires higher temperatures of more than 800°C (1,470°F).

Alkenes (hydrocarbons with at least one double bond) are produced as a result of cracking.

Shorter-chain alkanes, with no double bonds, are also produced.

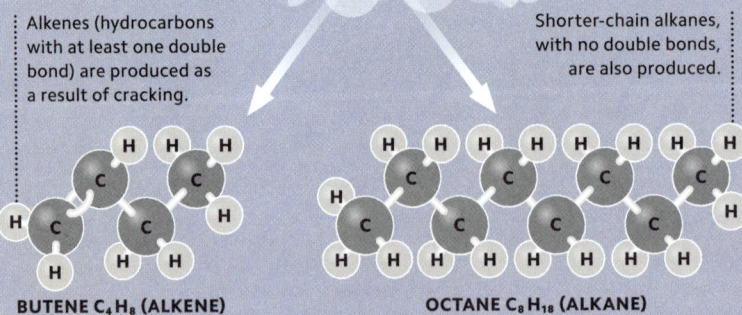

BUTENE C_4H_8 (ALKENE)   **OCTANE C_8H_{18} (ALKANE)**

100 | CRACKING ALKANES TO ALKENES

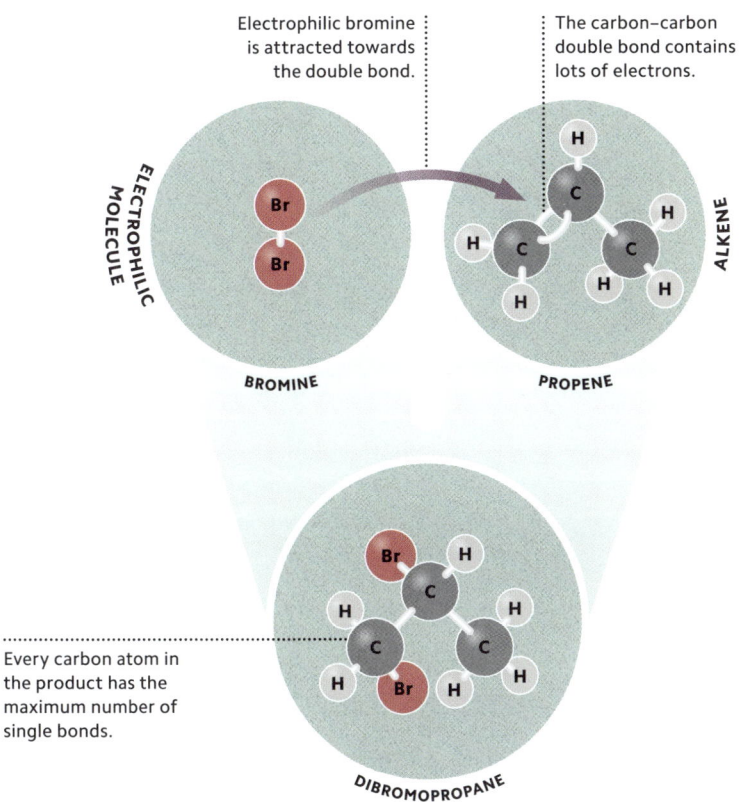

Every carbon atom in the product has the maximum number of single bonds.

REACTIVE HYDROCARBONS

The high concentration of electrons in an alkene double bond makes these hydrocarbons extremely reactive, particularly towards electron-hungry electrophilic molecules, such as bromine (Br_2). By breaking one of the two bonds between the alkene carbon atoms, each carbon atom can form a new bond with an atom from the electrophile. Overall, a new molecule is added across the alkene, converting the unsaturated alkene into a saturated compound (unable to add more single carbon bonds) that does not contain a double bond.

ALKENE REACTIONS

CHAIN BUILDING: Individual units of amino acids form a protein chain, called a polypeptide.

REPEATING PATTERNS: The polypeptide twists and folds to form either a helix structure or a pleated sheet.

3D FOLDING: The twisted or folded polypeptide chain can further form complicated 3D shapes.

COMBINING CHAINS: Some larger proteins are made up of several polypeptide chains assembled together.

Amino acid
Peptide bond

BETA SHEET
ALPHA HELIX

PRIMARY STRUCTURE | **SECONDARY STRUCTURE** | **TERTIARY STRUCTURE** | **QUATERNARY STRUCTURE**

BUILDING BLOCKS OF LIFE

Proteins are made of long chains of smaller molecules, called amino acids, which are chemically bonded together and folded up into a specific shape. There are 20 different naturally occurring amino acids, and each contains an amine (NH_2) group, a carboxylic acid (COOH) group, and a unique chemical side chain. The exact sequence of the different amino acids in the chain determines the shape of the overall protein, which enables it to carry out a particular function in the body.

NATURE'S SWEET FUEL

Carbohydrates are a huge family of biomolecules, all made from simple sugars, known as monosaccharides, and some chemically bonded together into long chains, known as polysaccharides. Starchy carbohydrates, such as those found in potatoes, bread, and pasta, are the body's main source of energy. Enzymes in the digestive system break these large polysaccharides into individual molecules of glucose, which can be absorbed by the intestines. This glucose is then either used immediately for energy or stored in the form of a different carbohydrate, called glycogen, until the body needs energy.

MONOMERS

DIMERS

POLYMERS

MONOSACCHARIDES

Monosaccharides, such as glucose, fructose, and galactose, are single sugar units. These are usually found in fruits, honey, and some types of cheese.

DISACCHARIDES

Two sugar units bond together to form a disaccharide. Sucrose, or table sugar, maltose, found in wheat, and lactose, found in milk, are examples of disaccharides.

POLYSACCHARIDES

Starch, glycogen, and cellulose are polysaccharides – many sugar units bonded together. Foods such as rice are starchy because of polysaccharides.

CARBOHYDRATES | 103

BIOLOGY'S BARRIERS

Lipids are biomolecules containing carbon, hydrogen, and oxygen that do not dissolve in water. These molecules are more familiar in the form of fats, technically termed triglycerides, found in foods such as butter, red meat, and oily fish. Triglycerides have three long hydrocarbon chains bonded to a single molecule of glycerol. The molecule is hydrophobic, meaning it repels water. Another group of lipids, called phospholipids, swap one hydrocarbon chain for a phosphate group, which is attracted to water, meaning it is hydrophilic. This combination of hydrophilic and hydrophobic properties means the molecules can interact with water.

PHOSPHOLIPID BILAYER: The phospholipid molecules are arranged in two layers, called a bilayer, so all the heads are next to the water and all the tails are protected on the inside.

Water is attracted to the bilayer, but cannot easily pass through it.

OUTSIDE CELL

CELL MEMBRANE

INSIDE CELL

The hydrocarbon tail repels water.

The phosphate head is strongly attracted to water.

Protective cell membrane
Phospholipids form the membranes of all cells in the body. The membrane acts as a protective barrier keeping the cell components in and harmful substances out.

NATURE'S PERFECT CATALYSTS

Enzymes are biological catalysts (see p.70) that ensure key reactions in living organisms happen fast enough to sustain life. They are made from proteins, folded to create a specific 3D shape that matches the shape of target molecules, also called substrates. When molecules enter the active site, they form an enzyme–substrate complex, which allows the enzyme to convert them into a product. As the product is a different shape from the substrate, it no longer fits in the active site and the enzyme releases the molecule, ready to catalyse another reaction.

Metabolic reactions
All enzyme-controlled metabolic reactions either build small molecules up in anabolic processes or break larger molecules down in catabolic processes.

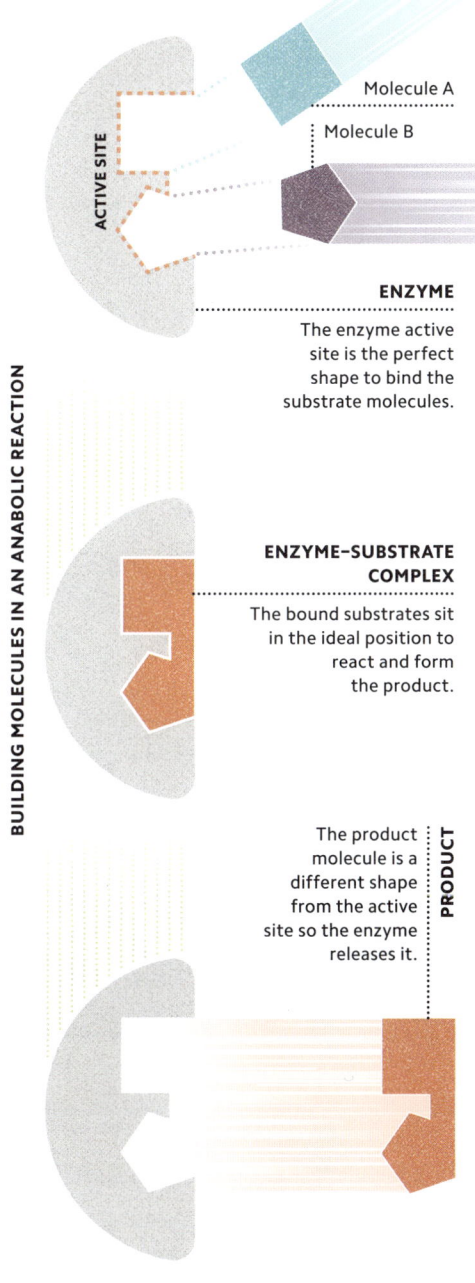

BUILDING MOLECULES IN AN ANABOLIC REACTION

ACTIVE SITE

Molecule A
Molecule B

ENZYME
The enzyme active site is the perfect shape to bind the substrate molecules.

ENZYME–SUBSTRATE COMPLEX
The bound substrates sit in the ideal position to react and form the product.

PRODUCT
The product molecule is a different shape from the active site so the enzyme releases it.

ENZYMES | 105

LIFE'S GENETIC SPIRAL

All the genetic information for an organism is held within its deoxyribonucleic acid (DNA). This molecule provides the instructions for how cells should produce the essential biomolecules to create life. DNA has a double helix structure, like a twisted ladder, formed from individual units called nucleotides. Each nucleotide contains a sugar molecule, a phosphate group, and one of four organic bases – cytosine, guanine, adenine, and thymine. Hydrogen bonds between the bases on different chains create the ladderlike structure, while their order encodes the genetic information.

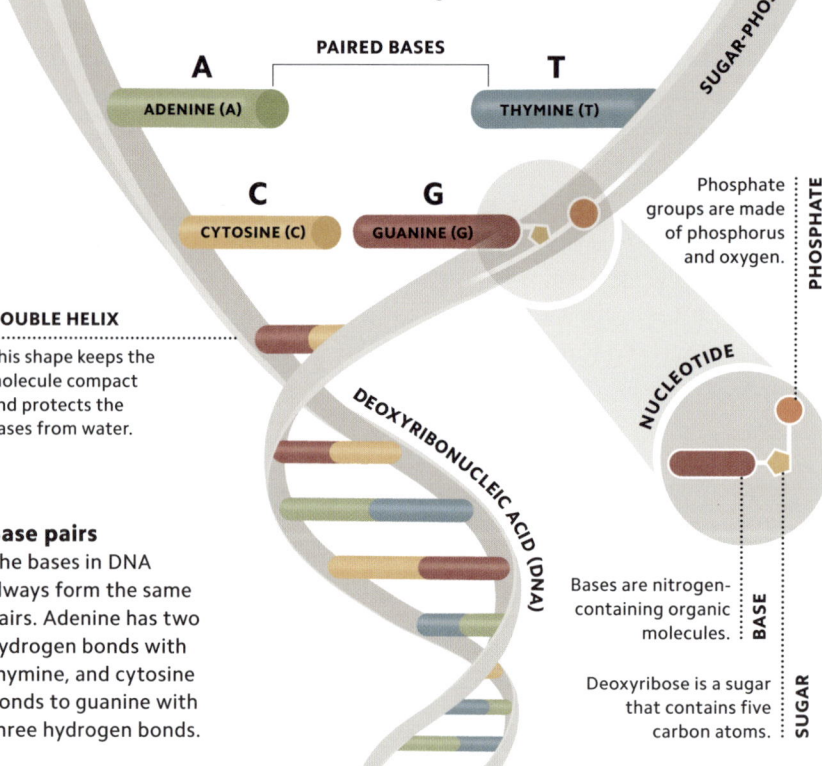

PAIRED BASES

- ADENINE (A)
- THYMINE (T)
- CYTOSINE (C)
- GUANINE (G)

SUGAR-PHOSPHATE BACKBONE

Phosphate groups are made of phosphorus and oxygen.

PHOSPHATE

DEOXYRIBONUCLEIC ACID (DNA)

NUCLEOTIDE

Bases are nitrogen-containing organic molecules.

BASE

Deoxyribose is a sugar that contains five carbon atoms.

SUGAR

DOUBLE HELIX
This shape keeps the molecule compact and protects the bases from water.

Base pairs
The bases in DNA always form the same pairs. Adenine has two hydrogen bonds with thymine, and cytosine bonds to guanine with three hydrogen bonds.

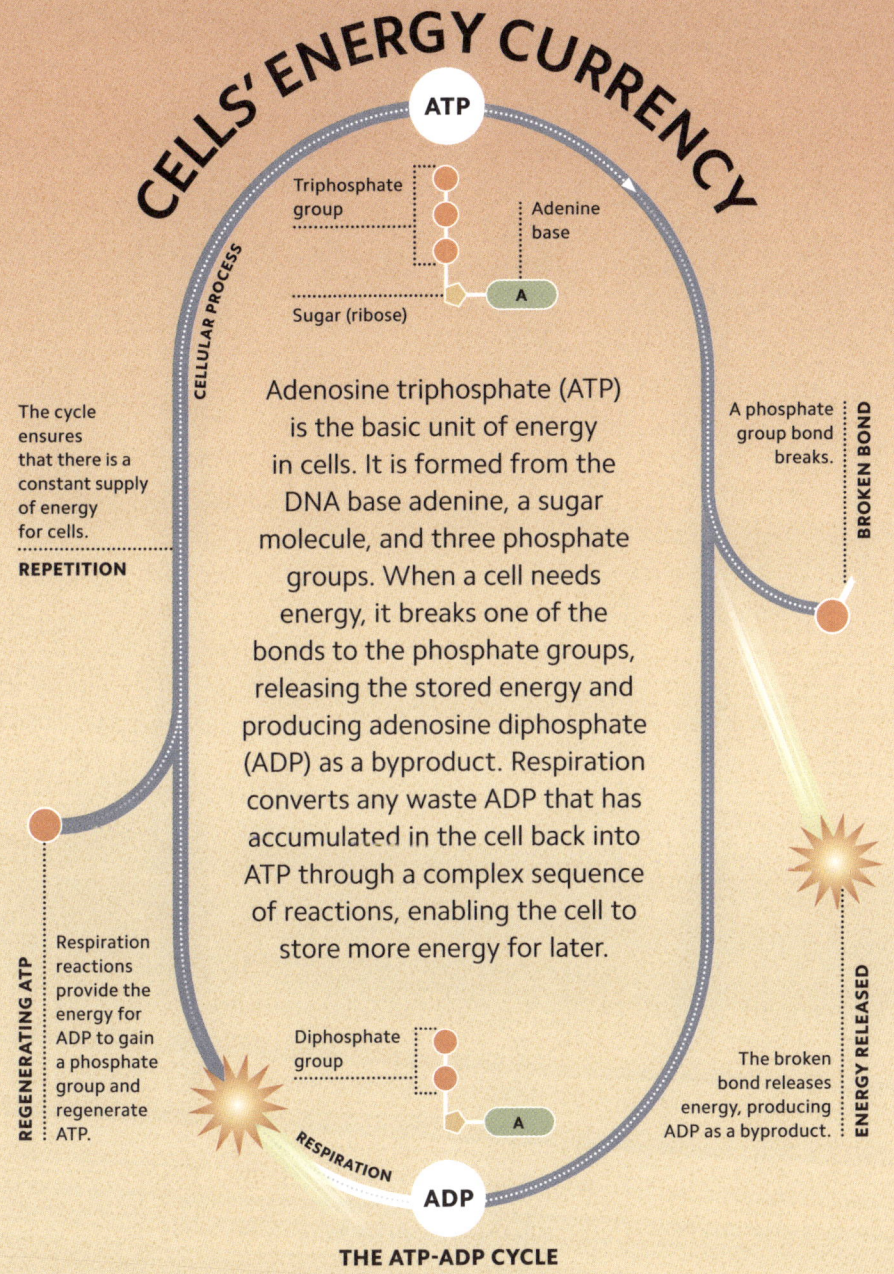

Oxygen's carrier
Cells require a continual supply of oxygen in order to function, and haemoglobin is a molecule specialized for oxygen transport.

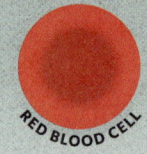
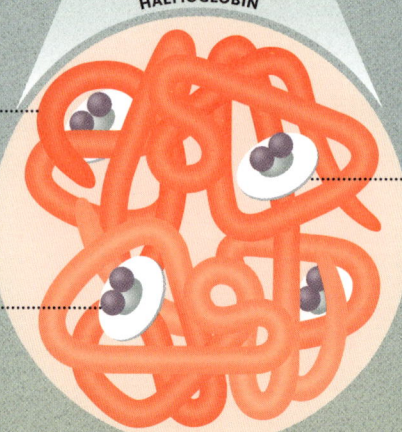

GLOBIN CHAIN
Four globin chains, which are ribbonlike chains of protein, are contained within haemoglobin.

OXYGEN
When collected, one oxygen (O_2) molecule binds to the iron within the haem group.

HAEM GROUP
Each globin chain combines with a haem molecule, which consists of a porphyrin ring and iron centre.

TRANSPORTING OXYGEN

Haemoglobin is the protein responsible for transporting oxygen around the body and is what gives blood its red colour. It contains a haem: a ferrous iron (Fe^{2+}) ion in the centre of a bowl-shaped molecule called a porphyrin ring. Four polypeptide chains called globins are each attached to a haem. In an oxygen-rich environment like the lungs, each haem binds to an oxygen molecule. The haemoglobin then carries that through the blood until it reaches an oxygen-poor environment, such as working muscle, where it releases the oxygen and returns to the lungs to pick up more.

CHEMICAL SIGNALLERS

A hormone is a type of chemical messenger that carries signals between different parts of the body to regulate essential biological processes, such as growth, reproduction, and metabolism. These molecules are usually produced by specialized organs called glands, which release hormones into the bloodstream. When a hormone reaches a specific receptor protein, it binds to the receptor to trigger a particular effect elsewhere in the body.

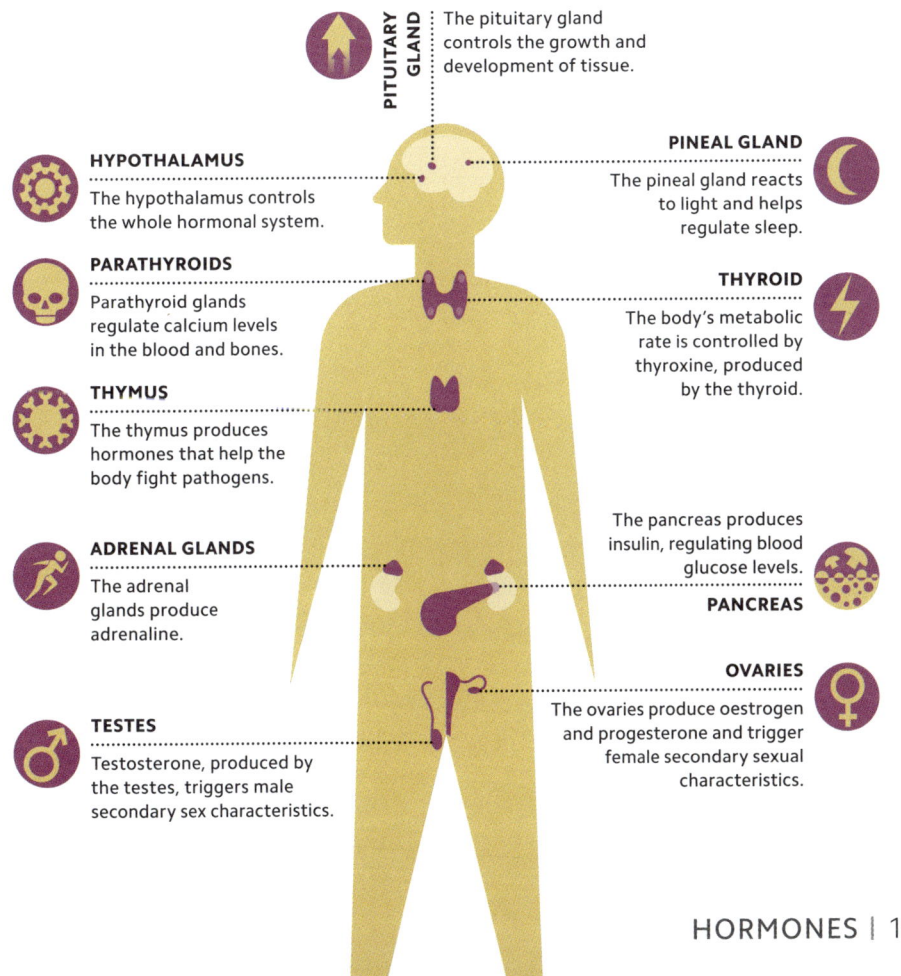

PITUITARY GLAND
The pituitary gland controls the growth and development of tissue.

HYPOTHALAMUS
The hypothalamus controls the whole hormonal system.

PARATHYROIDS
Parathyroid glands regulate calcium levels in the blood and bones.

THYMUS
The thymus produces hormones that help the body fight pathogens.

ADRENAL GLANDS
The adrenal glands produce adrenaline.

TESTES
Testosterone, produced by the testes, triggers male secondary sex characteristics.

PINEAL GLAND
The pineal gland reacts to light and helps regulate sleep.

THYROID
The body's metabolic rate is controlled by thyroxine, produced by the thyroid.

PANCREAS
The pancreas produces insulin, regulating blood glucose levels.

OVARIES
The ovaries produce oestrogen and progesterone and trigger female secondary sexual characteristics.

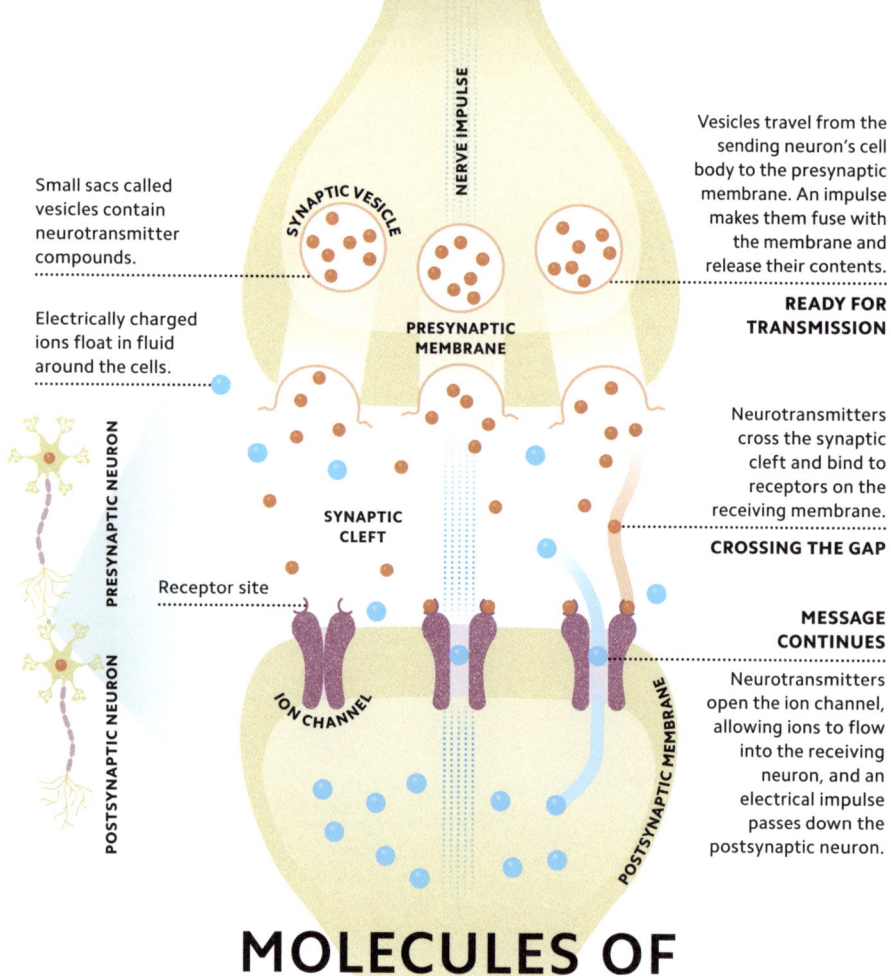

Small sacs called vesicles contain neurotransmitter compounds.

Electrically charged ions float in fluid around the cells.

Receptor site

SYNAPTIC VESICLE
NERVE IMPULSE
PRESYNAPTIC MEMBRANE
SYNAPTIC CLEFT
ION CHANNEL
POSTSYNAPTIC MEMBRANE
PRESYNAPTIC NEURON
POSTSYNAPTIC NEURON

Vesicles travel from the sending neuron's cell body to the presynaptic membrane. An impulse makes them fuse with the membrane and release their contents.
READY FOR TRANSMISSION

Neurotransmitters cross the synaptic cleft and bind to receptors on the receiving membrane.
CROSSING THE GAP

MESSAGE CONTINUES
Neurotransmitters open the ion channel, allowing ions to flow into the receiving neuron, and an electrical impulse passes down the postsynaptic neuron.

MOLECULES OF CONSCIOUSNESS

At the end of every nerve cell, or neuron, is a small gap to the next cell, called a synaptic cleft. Electric nerve signals cannot cross this gap, so the neurons use chemical messengers, called neurotransmitters, to pass the signal on. When the signal reaches the end of a neuron, it triggers tiny pockets (synaptic vesicles) to release neurotransmitters into the synaptic cleft. They diffuse across the gap and bind to receptors on the next neuron, stimulating it to pass on the signal.

CLEANING UP

Surfactants are chemical compounds that reduce the surface tension of liquids, allowing substances to mix together that would otherwise be unable to – such as oil and water. Their chemical structure contains two parts: one attracted to water, the other repelled by it. Many household products rely on surfactant molecules. For example, laundry detergents use this combination of water-loving and water-hating properties to pull greasy stains from clothing and help them dissolve in water.

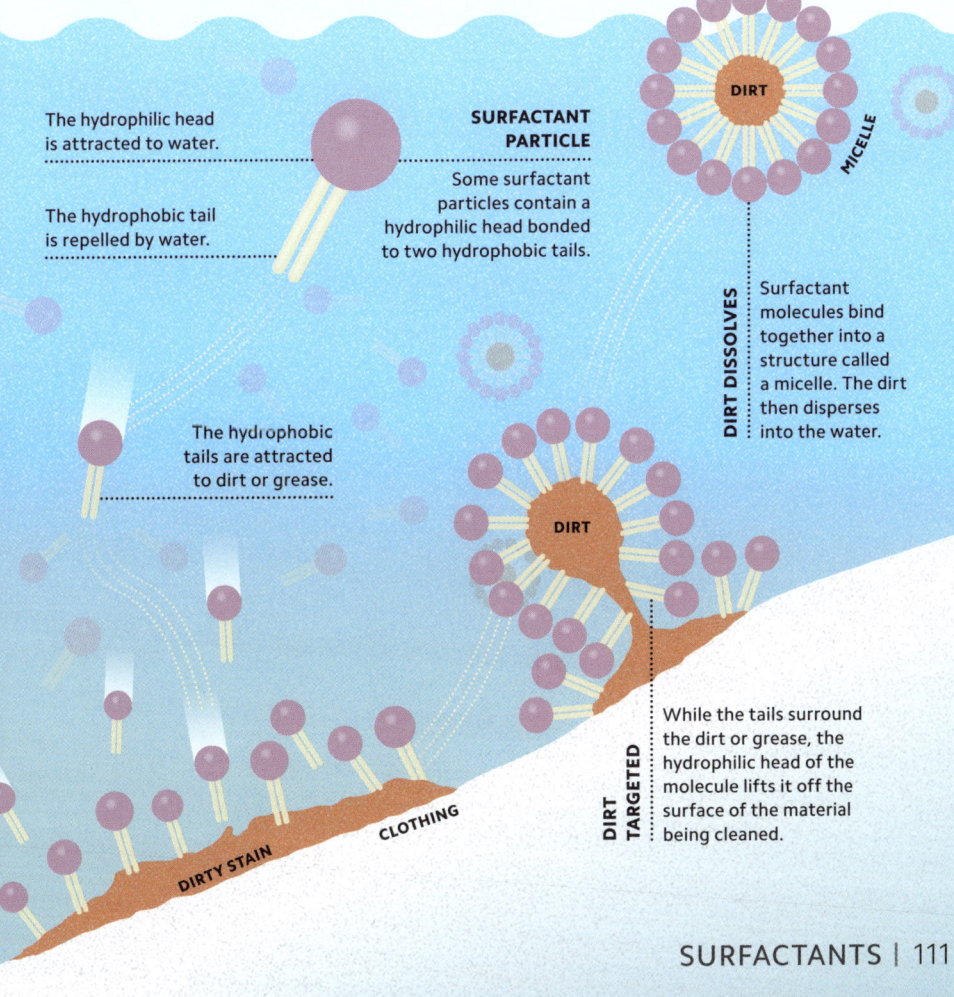

LIVING FACTORIES

Living organisms rely on a vast array of complex molecules, from the proteins that make up enzymes to the nucleobases that encode the information in DNA. Manufacturing these compounds is no simple matter and requires a complicated production line of chemical reactions, known as a biosynthetic pathway. Each step along the pathway effects a small change, adding or changing a chemical group to slowly transform simple molecules into something more intricate. Many of the steps are controlled by enzymes, which ensure that each reaction happens quickly and correctly.

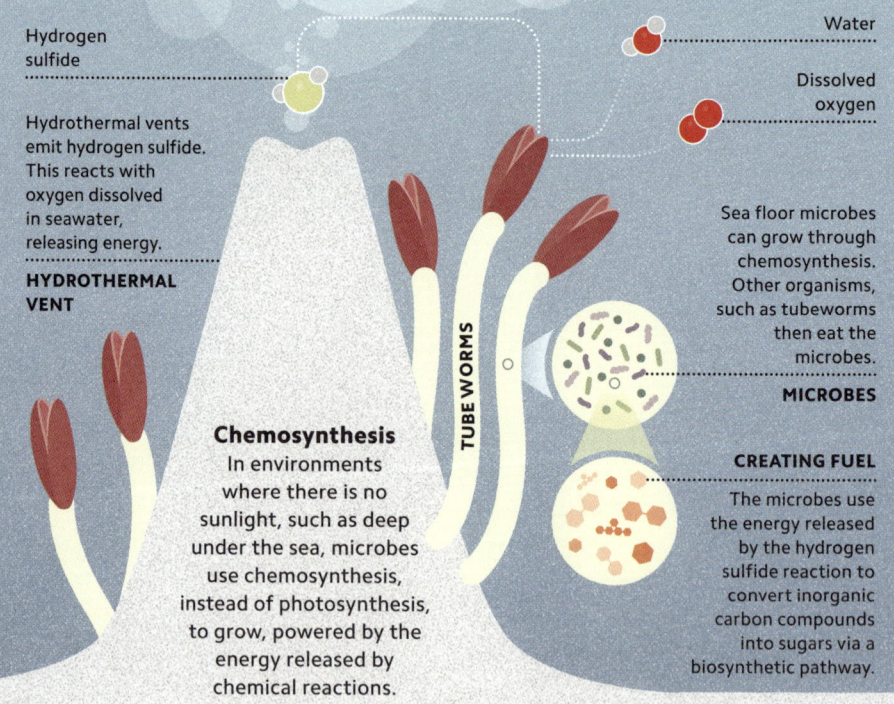

Hydrogen sulfide

Hydrothermal vents emit hydrogen sulfide. This reacts with oxygen dissolved in seawater, releasing energy.

HYDROTHERMAL VENT

Chemosynthesis
In environments where there is no sunlight, such as deep under the sea, microbes use chemosynthesis, instead of photosynthesis, to grow, powered by the energy released by chemical reactions.

TUBE WORMS

Water

Dissolved oxygen

Sea floor microbes can grow through chemosynthesis. Other organisms, such as tubeworms then eat the microbes.

MICROBES

CREATING FUEL

The microbes use the energy released by the hydrogen sulfide reaction to convert inorganic carbon compounds into sugars via a biosynthetic pathway.

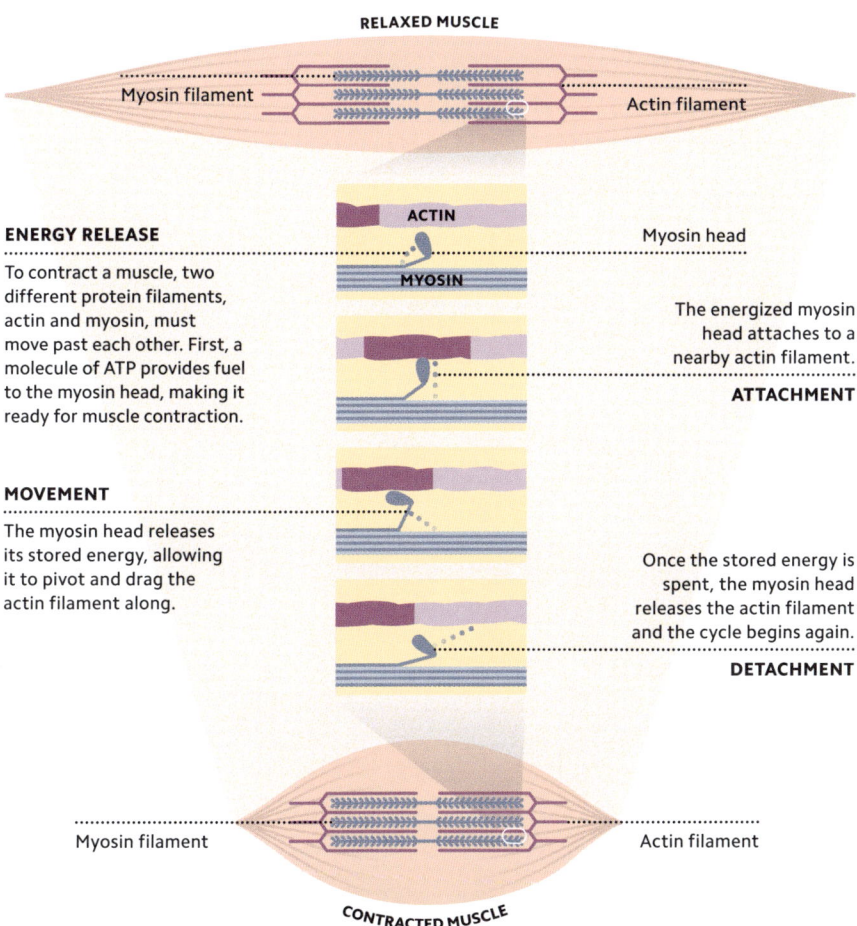

TINY GEARS IN ACTION

Just like a conventional machine, molecular machines contain moving parts that work together to perform a particular task. A chemical reaction provides the fuel, changing the shape of one molecular component to trigger the movement of another. In the human body, hundreds of these "nanomachines" support life-sustaining processes, such as muscle contraction and cell division.

NATUR
CHEMI

AL
STRY

Chemistry holds the key to our existence, on the only planet where we know there is life. Through the availability of water, the development of photosynthesis, and the evolution of a breathable atmosphere, the early Earth fostered emerging life. Over billions of years since, plants and animals of all kinds have grown and flourished. But the times are changing, and with them the planet's resilience. Chemistry can describe how modern civilizations have altered the climate and the environment. It can also suggest solutions that may alleviate the impacts we are facing as a species – if action is taken to implement them.

THE MOLECULE OF LIFE

Water's unusual properties make it unlike any other liquid on Earth. It easily dissolves other substances and is crucial for life-sustaining chemical reactions. In liquid form it has high surface tension, which allows it to form bubbles, and in solid form it becomes less dense, which allows it to float. These properties arise from water's simple chemical structure. One slightly negatively charged oxygen atom is bonded to two slightly positively charged hydrogen atoms to form a V-shaped molecule. The water molecules can "stick" together through hydrogen bonds, which result from the oppositely charged atoms.

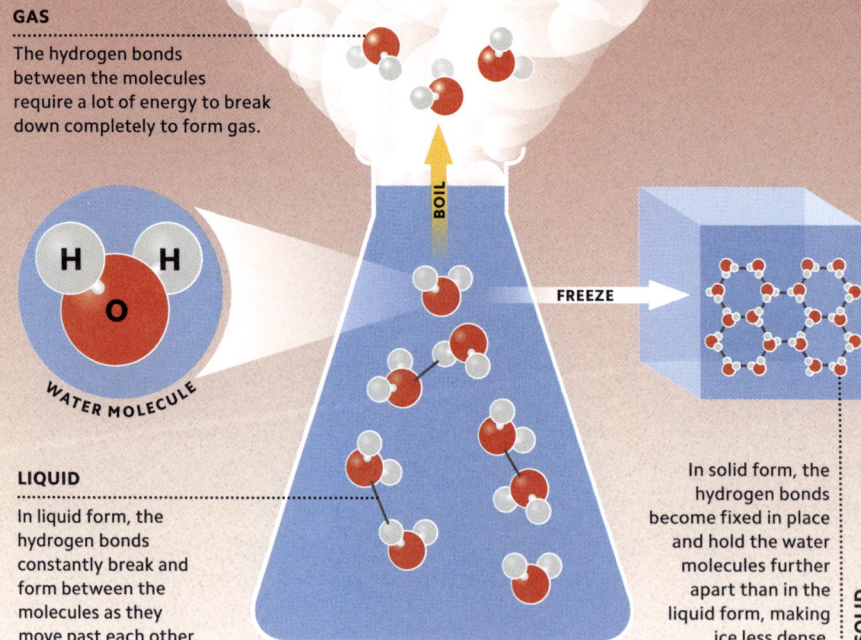

GAS
The hydrogen bonds between the molecules require a lot of energy to break down completely to form gas.

LIQUID
In liquid form, the hydrogen bonds constantly break and form between the molecules as they move past each other.

SOLID
In solid form, the hydrogen bonds become fixed in place and hold the water molecules further apart than in the liquid form, making ice less dense.

THE PROPERTIES OF WATER

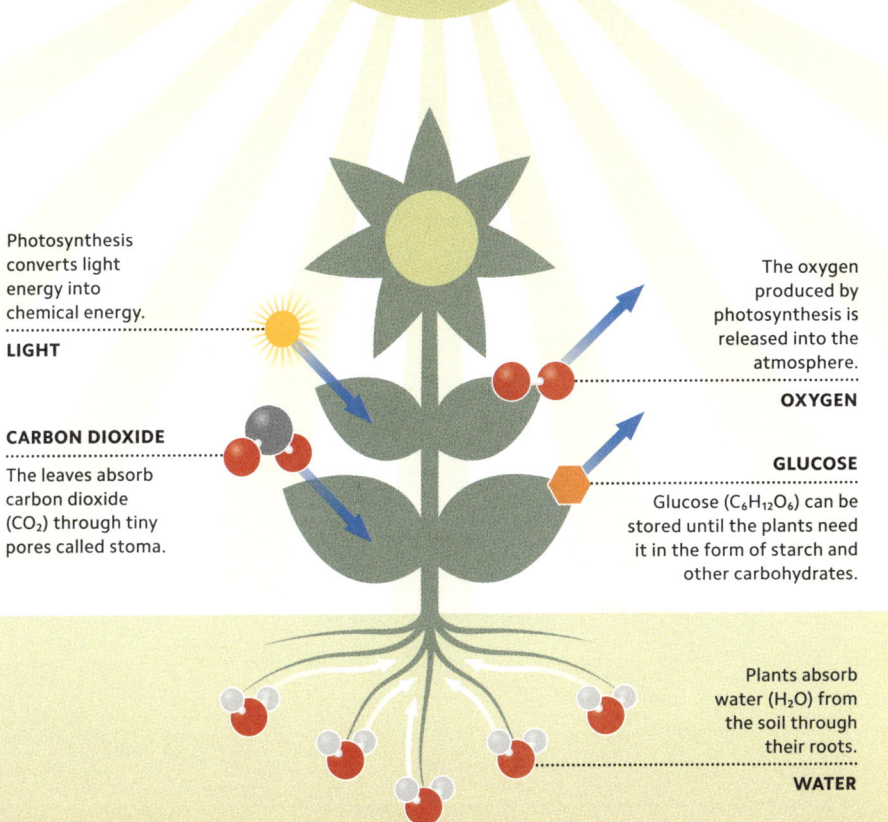

TURNING LIGHT INTO LEAVES

Plants use energy from the Sun to make their own food from gases in the atmosphere. This complex chemical reaction, known as photosynthesis, takes place in saclike structures called chloroplasts, which contain a special green pigment called chlorophyll. During the day, this pigment absorbs sunlight, powering a chain of chemical reactions that combine carbon dioxide and water to produce oxygen and glucose, a type of sugar. When the plant needs energy, this glucose is used as fuel in another chemical reaction called respiration.

PHOTOSYNTHESIS

> **Earth's early atmosphere was probably mostly CO_2, similar to the atmospheres of Mars and Venus today.**

Most scientists believe Earth's atmosphere formed shortly after the planet itself, following a spate of violent volcanic activity. While it is impossible to know its precise composition at the time, it was probably a noxious cocktail of volcanic gases – around 95 per cent carbon dioxide, with small amounts of hydrogen sulfide, methane, nitrogen, and water vapour, but no oxygen. The high levels of carbon dioxide trapped heat in the atmosphere, making the planet much hotter than it is today, with some estimates placing the surface temperature at around 70 °C (158 °F).

HYDROGEN SULFIDE (H_2S) **WATER (H_2O)** **CARBON DIOXIDE (CO_2)** **METHANE (CH_4)** **NITROGEN (N_2)**

FROM HELL TO EARTH

BREATHING CHEMISTRY

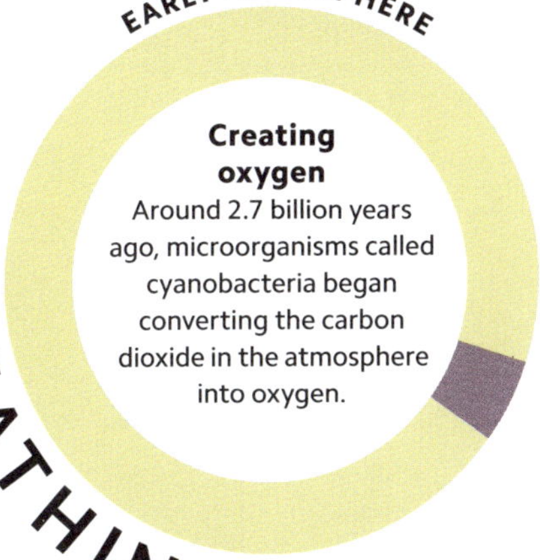

EARLY ATMOSPHERE

KEY
- Nitrogen (N$_2$)
- Oxygen (O$_2$)
- Argon (Ar)
- Carbon dioxide (CO$_2$)
- Other gases
- Trace gases

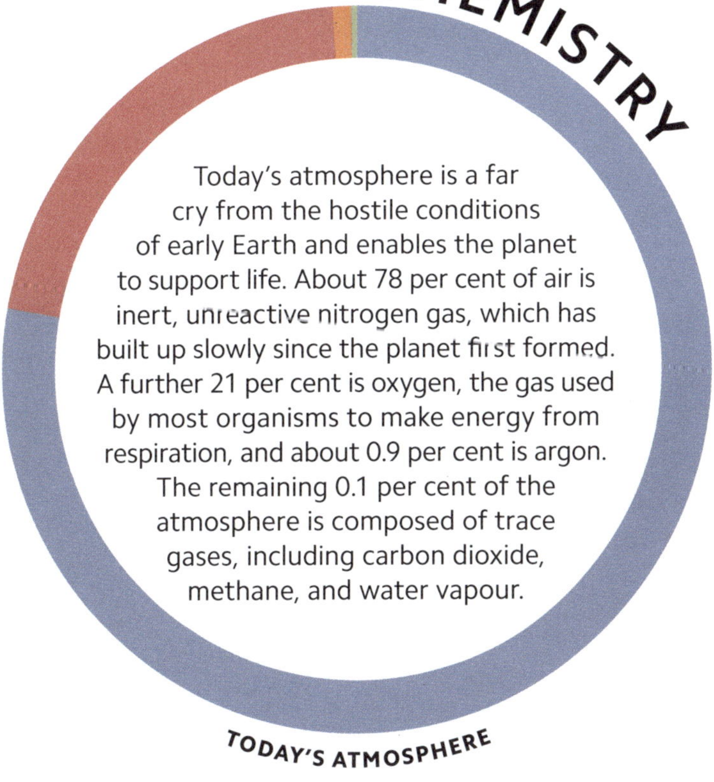

Creating oxygen
Around 2.7 billion years ago, microorganisms called cyanobacteria began converting the carbon dioxide in the atmosphere into oxygen.

Today's atmosphere is a far cry from the hostile conditions of early Earth and enables the planet to support life. About 78 per cent of air is inert, unreactive nitrogen gas, which has built up slowly since the planet first formed. A further 21 per cent is oxygen, the gas used by most organisms to make energy from respiration, and about 0.9 per cent is argon. The remaining 0.1 per cent of the atmosphere is composed of trace gases, including carbon dioxide, methane, and water vapour.

TODAY'S ATMOSPHERE

MODERN ATMOSPHERIC COMPOSITION

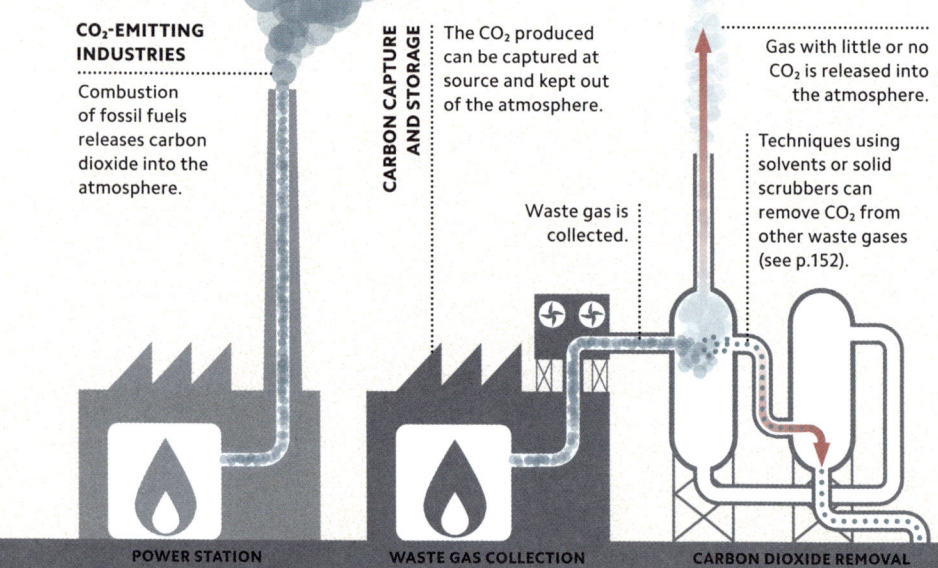

CO₂-EMITTING INDUSTRIES
Combustion of fossil fuels releases carbon dioxide into the atmosphere.

CARBON CAPTURE AND STORAGE
The CO₂ produced can be captured at source and kept out of the atmosphere.

Waste gas is collected.

Gas with little or no CO₂ is released into the atmosphere.

Techniques using solvents or solid scrubbers can remove CO₂ from other waste gases (see p.152).

POWER STATION

WASTE GAS COLLECTION

CARBON DIOXIDE REMOVAL

Removing carbon
Some industrial processes now capture the carbon dioxide produced from the burning of fossil fuels and transport it to another location for storage underground. This is known as carbon capture and storage.

The carbon dioxide is transported underground to a processing facility.

CLIMATE CULPRITS

TRAPPING

Fossil fuels are an excellent source of energy, but burning them releases carbon that has been trapped deep underground for millions of years. Coal and gas power stations produce huge amounts of carbon dioxide – a potent greenhouse gas – slowly increasing its concentration in Earth's atmosphere. Carbon dioxide, and other greenhouse gases such as methane and nitrous oxide, are responsible for global warming. They trap some heat energy from the Sun in Earth's atmosphere, making the planet warmer than it would be otherwise.

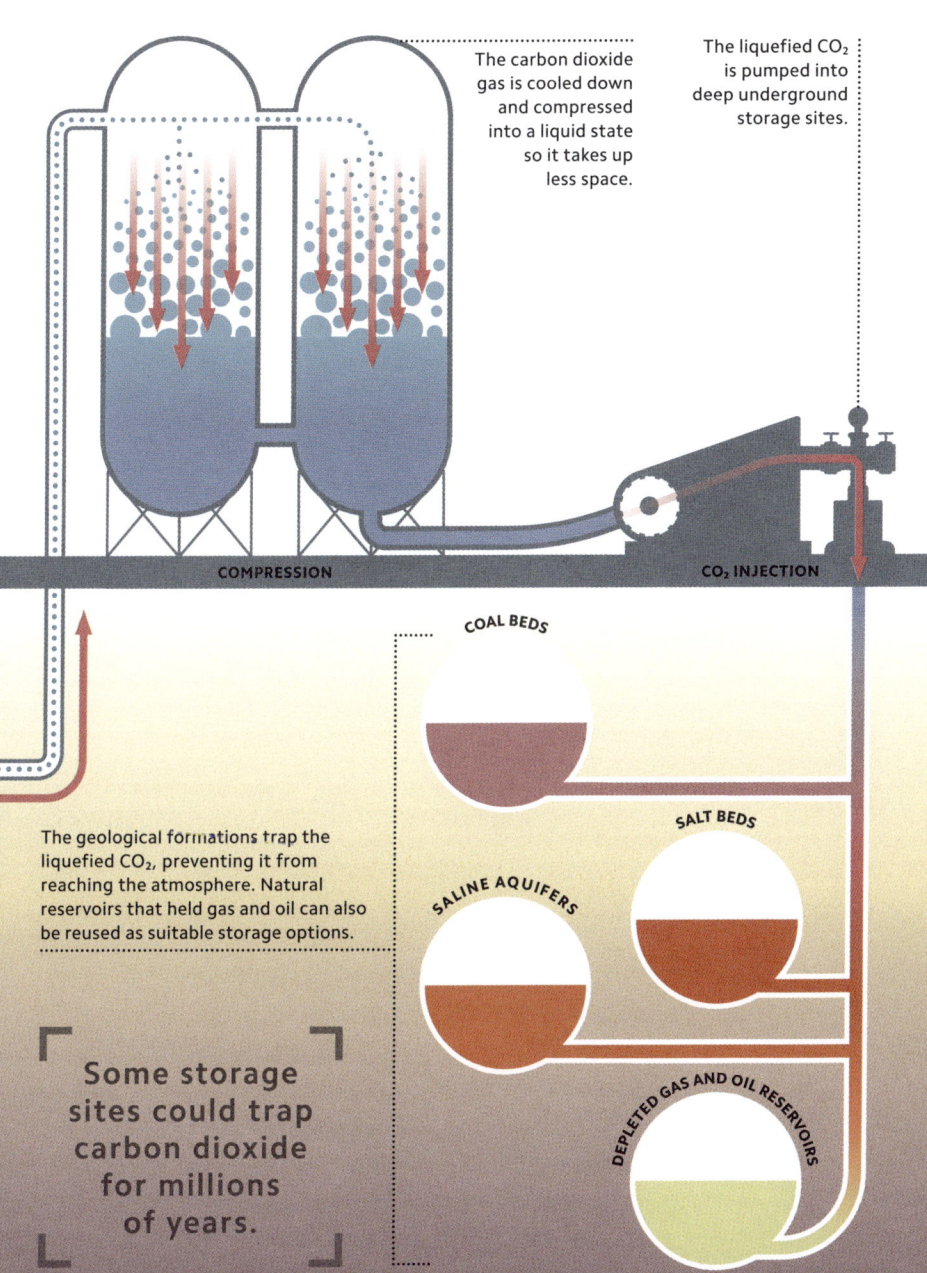

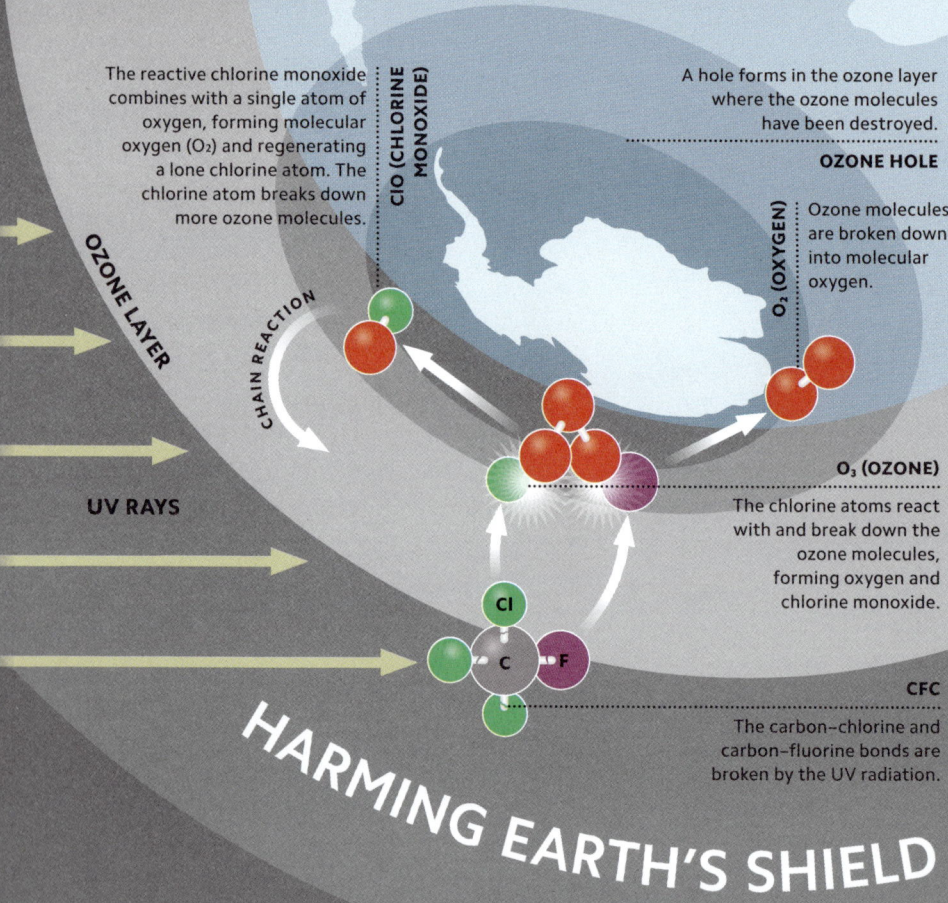

HARMING EARTH'S SHIELD

The ozone layer acts as a protective barrier around Earth, shielding the planet's surface from the Sun's strongest cancer-causing ultraviolet (UV) radiation. However, in 1985, scientists discovered that it had become damaged by a group of human-made chemicals called chlorofluorocarbons (CFCs) and a hole had formed over the Antarctic. UV radiation breaks the carbon–chlorine and carbon–fluorine bonds in CFCs, producing reactive chlorine atoms that destroy the protective ozone molecules. A global agreement in 1987 phased out the use of these chemicals and the ozone layer has since begun to recover. It is expected to completely heal by the middle of the 21st century.

Although a seemingly vast expanse of nothingness, outer space is teeming with chemical activity. The lighter chemical elements produced in the Big Bang – hydrogen (H), carbon (C), and oxygen (O) – combine and break apart to form simple organic molecules like ethanol (C_2H_6O), methane (CH_4), and formaldehyde (CH_2O). Scientists have already identified more than 250 different molecules in space. Understanding how these basic molecules are made helps researchers build a picture of how the Universe began and could even reveal the origin of life.

MOLECULES BEYOND EARTH

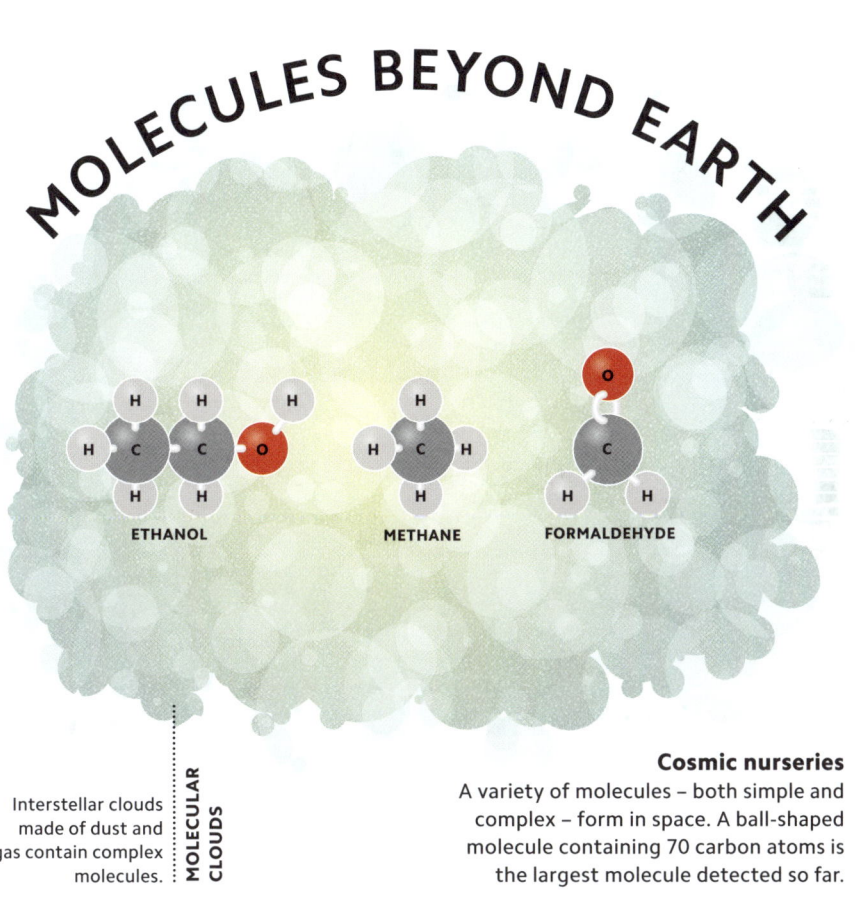

ETHANOL

METHANE

FORMALDEHYDE

Interstellar clouds made of dust and gas contain complex molecules.

MOLECULAR CLOUDS

Cosmic nurseries
A variety of molecules – both simple and complex – form in space. A ball-shaped molecule containing 70 carbon atoms is the largest molecule detected so far.

CHEMICALS IN SPACE | 123

INDUS
CHEMI

INDUSTRIAL CHEMISTRY

The age of industrial chemistry surged ahead with the extraction and processing of the fuels that powered the industrial revolution. Since then, chemistry has been conducted on an unprecedented scale across the world, in pursuit of plastics, pharmaceuticals, agricultural chemicals, detergents, and more. Transformative discoveries of fertilizers and medicines have been among the outcomes of modern industrial chemistry. So too have the techniques to unravel our genetic code and build astonishing nanoscale technologies. Yet the impacts of industrial processes are an experiment affecting the entire globe, whose outcomes we are only starting to acknowledge and address.

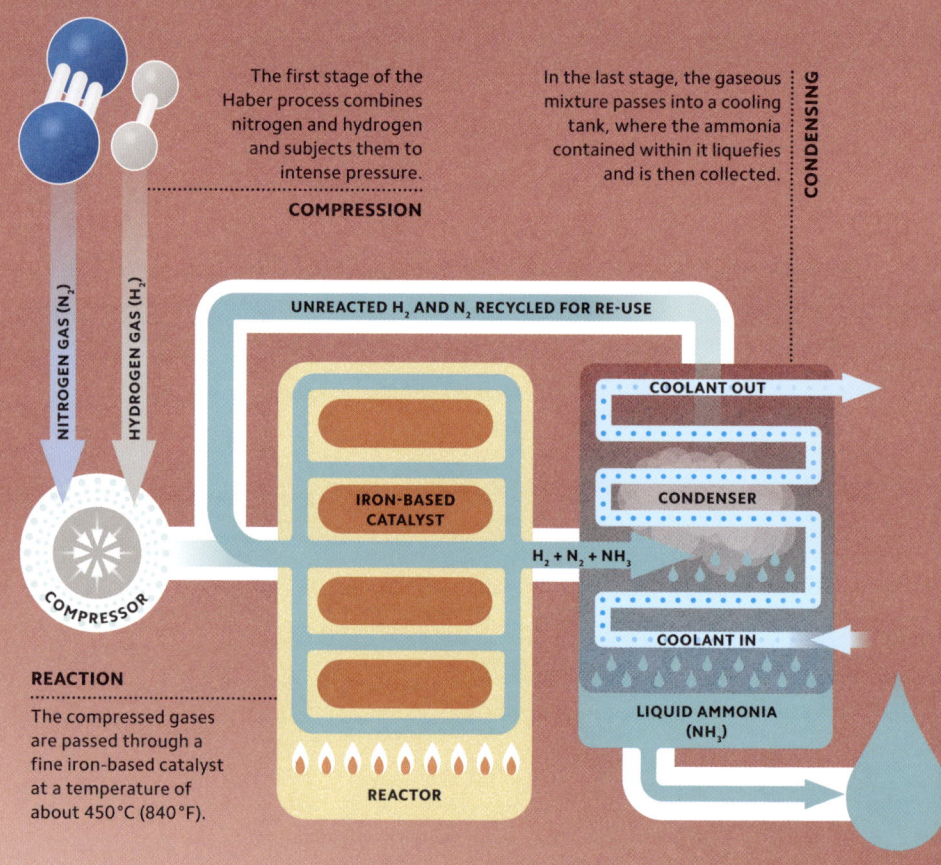

MAKING AMMONIA

Ammonia (NH_3) is an important industrial chemical, made from hydrogen (H_2) and nitrogen (N_2). Both substances are readily available, but nitrogen has a strong triple covalent bond, making it unreactive, and so it does not easily combine with hydrogen. In the early 20th century, Fritz Haber developed an efficient method of ammonia production. The Haber process uses pressure and a catalyst (see p.70) to speed up the reaction between the two reactants and produce viable amounts of ammonia for use in fertilizers, explosives, and dyes.

POWERING PLANTS

Plants need nitrogen to grow. Although Earth's atmosphere is almost 80 per cent nitrogen, it exists in a form that vegetation cannot absorb directly. The nitrogen molecule needs to be broken down first. In nature, microorganisms in the soil do this in a process called nitrogen fixation. Formed by chemical reactions between ammonia and acids, fertilizers are salts that dissolve in water, providing nitrogen in a form that plant roots can absorb.

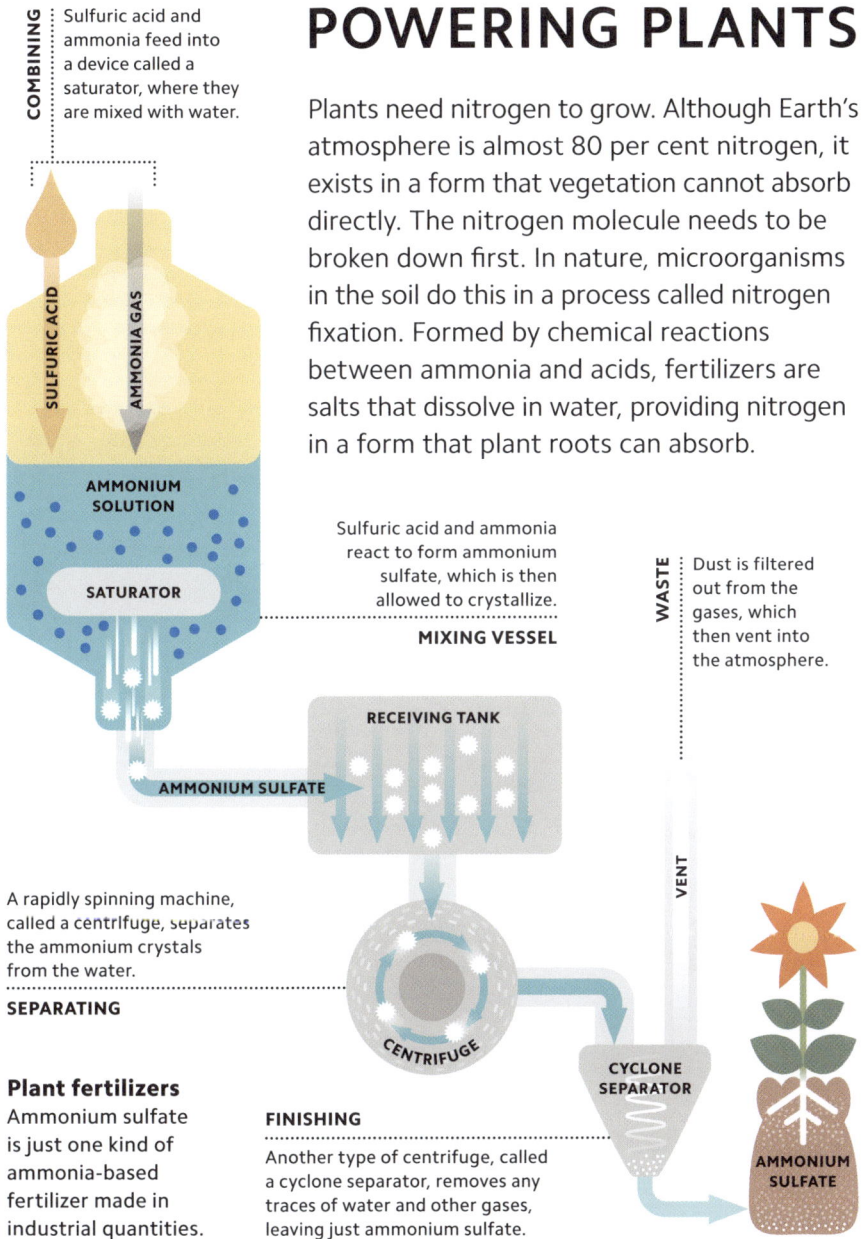

COMBINING
Sulfuric acid and ammonia feed into a device called a saturator, where they are mixed with water.

MIXING VESSEL
Sulfuric acid and ammonia react to form ammonium sulfate, which is then allowed to crystallize.

WASTE
Dust is filtered out from the gases, which then vent into the atmosphere.

SEPARATING
A rapidly spinning machine, called a centrifuge, separates the ammonium crystals from the water.

FINISHING
Another type of centrifuge, called a cyclone separator, removes any traces of water and other gases, leaving just ammonium sulfate.

Plant fertilizers
Ammonium sulfate is just one kind of ammonia-based fertilizer made in industrial quantities.

MAKING FERTILIZERS | 127

MOULDABLE MARVELS

Plastics are synthetic polymers, large molecules composed of repeating smaller units called monomers (see pp.72–73). The exact chemical configuration of these units determines the plastic's final form. Some, like polyethylene and polypropylene, soften when heated and are easily reshaped; others, such as epoxy resin and polyurethane, do not melt at high temperatures and are used for insulation and fire-resistant coatings. The strong carbon–carbon bonds in plastics often prevent them from breaking down in the environment.

Polymer families
The different thermal properties of plastics are largely determined by the bonds between the polymer chains. The spatial arrangement of the atoms can also have an effect.

> It is thought that by the year 2050, the amount of plastic pollution in the ocean will outweigh the amount of fish.

THERMOSETS

Thermoset plastics are hard, brittle, and do not melt at high temperature. During manufacture, the material undergoes a process called curing, which forms cross-linking bonds between the different polymer chains, fixing them in place.

Strong cross-links prevent polymer chains from moving once set, even when heated.

THERMOPLASTICS

Thermoplastics make up around 80 per cent of all plastics. Unlike thermosets, they can be reshaped and remoulded many times by heating, making them easy to recycle. Their long polymer chains are held together by weak forces between molecules, with no cross-links connecting them.

Polymer chains are held together by weak forces that weaken further when heated.

AMORPHOUS

Amorphous plastics have a random and disorganized arrangement of polymer chains. This makes them easier to melt, but the reshaped material is more brittle.

SEMI-CRYSTALLINE

Semi-crystalline plastics have random areas where the polymer chains are organized, forming tiny crystalline areas. These plastics are durable, but break under sudden impact.

Structured region

Disordered region

PLASTICS | 129

CALLING TIME ON FOSSIL FUELS

Fossil fuels are so-named because they form from the buried remains of prehistoric organisms. Coal and methane originate from plants, whereas oil is largely from small sea creatures. These fuels are hydrocarbons, which are compounds that consist of only carbon and hydrogen, although they also contain some impurities. When methane, coal, and oil products such as petrol (gasoline) and diesel are burned as fuel, they release carbon dioxide (CO_2) into the atmosphere, contributing to climate change.

Global warming
Earth is warming at about 0.20 °C (0.36 °F) per decade, largely due to atmospheric CO_2. Getting, processing, and burning fossil fuels is greatly affecting the planet's climate. Chemists are helping develop viable cleaner alternatives (see opposite).

Biofuel
Biofuel production utilizes the ability of algae to turn carbon dioxide and water into energy-rich hydrocarbon molecules.

Vehicles can run on biofuels instead of petrol (gasoline) or traditional diesel.

Algae are tiny organisms that photosynthesize and absorb CO_2.

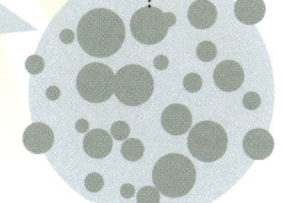

RELEASES CO_2

SUNLIGHT

CARBON-RICH OIL EXTRACTED FROM ALGAE

FUEL USED IN SAME WAY AS FOSSIL FUELS

CLEANER POWER

Alternative fuels are fuels that do not come from crude oil (see p.76). Examples include hydrogen (see p.132) and biofuels made from plants or algae. Although biofuels do release carbon dioxide (CO_2) when they are burned, plants and algae grown for biofuels absorb CO_2, so the overall increase is smaller. Biofuels produce less pollution than fossil fuels, can be a useful way to recycle materials such as waste cooking oil, and are renewable, meaning we can make more as needed.

REFINED INTO FUEL

Fuel can be distributed using standard fuel pumps.

The oil needs to be refined before it can be used.

ALTERNATIVE FUELS

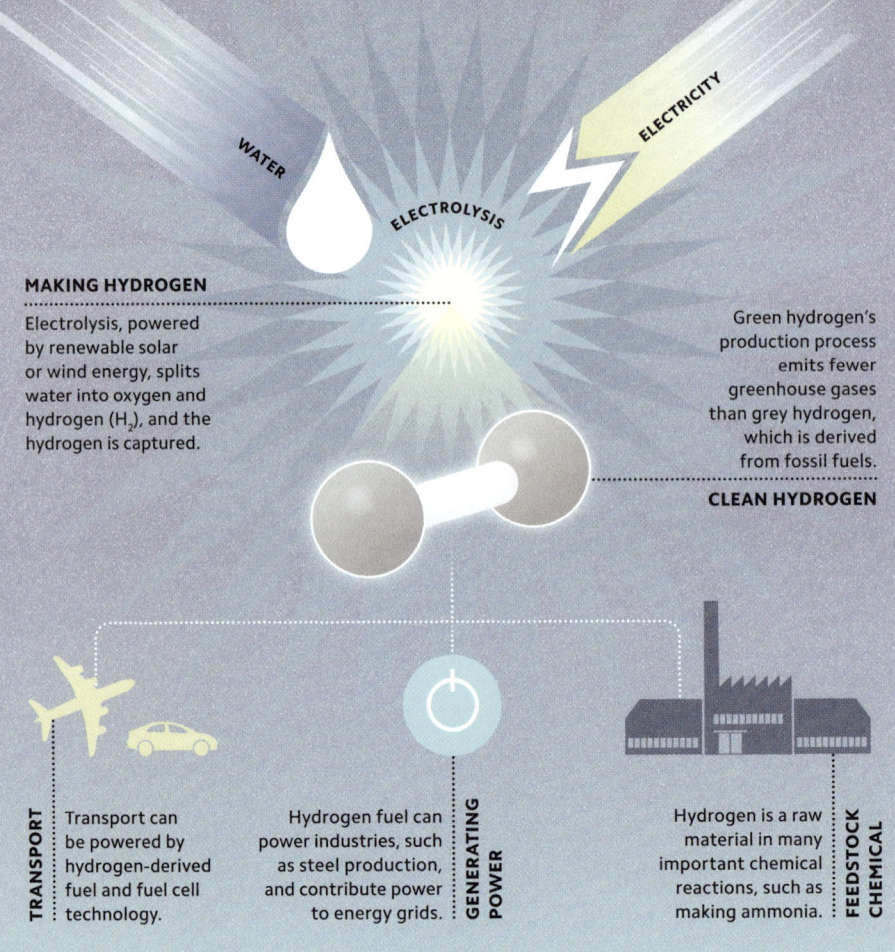

MAKING HYDROGEN

Electrolysis, powered by renewable solar or wind energy, splits water into oxygen and hydrogen (H_2), and the hydrogen is captured.

CLEAN HYDROGEN

Green hydrogen's production process emits fewer greenhouse gases than grey hydrogen, which is derived from fossil fuels.

TRANSPORT — Transport can be powered by hydrogen-derived fuel and fuel cell technology.

GENERATING POWER — Hydrogen fuel can power industries, such as steel production, and contribute power to energy grids.

FEEDSTOCK CHEMICAL — Hydrogen is a raw material in many important chemical reactions, such as making ammonia.

SPLITTING WATER

Green hydrogen is essential for a fossil fuel-free world. On Earth, hydrogen is typically combined with other elements, such as oxygen in water (H_2O). Industrial companies produce hydrogen gas (H_2) from fossil fuels, but to source green hydrogen, chemists can split water into hydrogen and oxygen using electrolysis (see p.69) and renewable energy. Using hydrogen as a fuel is beneficial for the environment because when it reacts with oxygen, it only produces water as a by-product ($2H_2 + O_2 \rightarrow 2H_2O$), so there is no pollution.

MOLECULAR RENEWAL

Current plastic recycling involves melting down old products and reshaping them into something new. However, this only works for thermosoftening plastics (see pp.128–29), which soften when heated, so a huge amount of plastic waste cannot yet be effectively recycled. Scientists are working on alternative recycling processes, including some that break the plastic's polymer chain into the individual monomer molecules through a chemical reaction known as depolymerization. But, because every polymer is built from different monomers, each plastic needs a different depolymerization reaction, which researchers must develop separately.

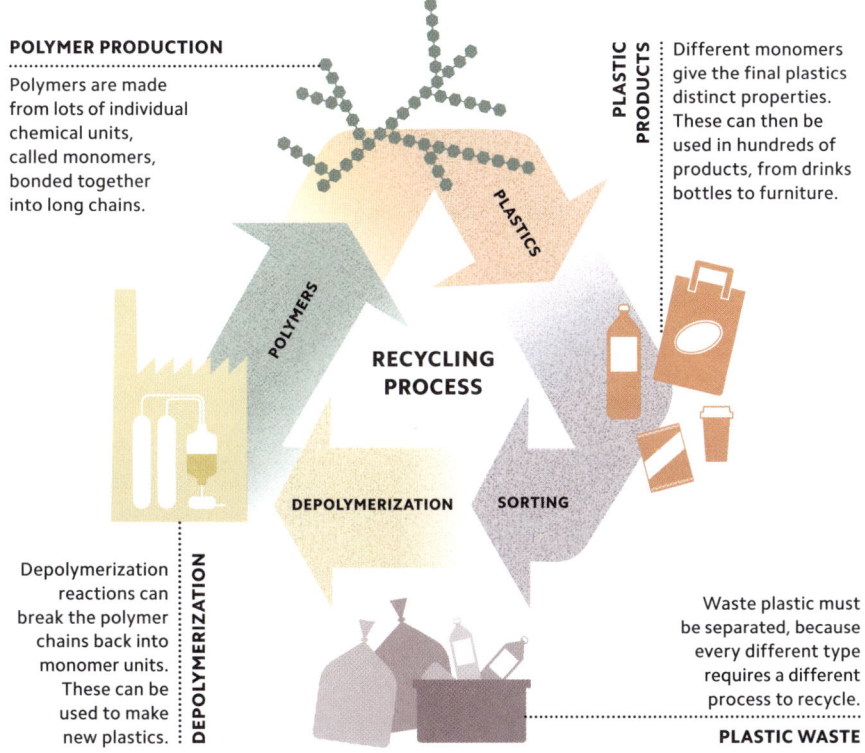

POLYMER PRODUCTION
Polymers are made from lots of individual chemical units, called monomers, bonded together into long chains.

PLASTIC PRODUCTS
Different monomers give the final plastics distinct properties. These can then be used in hundreds of products, from drinks bottles to furniture.

RECYCLING PROCESS
POLYMERS — PLASTICS — SORTING — DEPOLYMERIZATION

Depolymerization reactions can break the polymer chains back into monomer units. These can be used to make new plastics.

PLASTIC WASTE
Waste plastic must be separated, because every different type requires a different process to recycle.

CHEMISTRY VS. NATURE'S PESTS

In order to maximize their crop yields, farmers spray their fields with chemicals called herbicides and pesticides. These controlled substances protect the harvest from pests or competing weeds by inhibiting key enzymes, disrupting life-sustaining processes, or preventing cell division. They may either target a particular species or provide general protection. Although small amounts are usually sufficient to defend against infestation, these chemicals can travel through air, water, or soil and pollute the surrounding areas that were not meant to be targeted. As many pesticides have been linked to diseases in animals including humans, general pesticides are now regulated.

A toxic food chain
The increase in the concentration, and as a result the toxicity, of pesticides as they move up the food chain is known as bioaccumulation.

An aphid attacks a plant sprayed with insecticide and consumes 1 molecule of insecticide.

A spider eats 10 aphids and so consumes 10 molecules of insecticide.

A chicken eats 10 spiders and so consumes 100 molecules of insecticide.

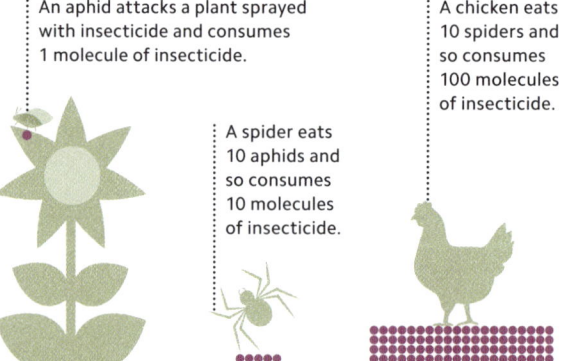

The average American eats the equivalent of 25–30 whole chickens each year, ingesting a possible 3,000 molecules of insecticide.

134 | HERBICIDES AND PESTICIDES

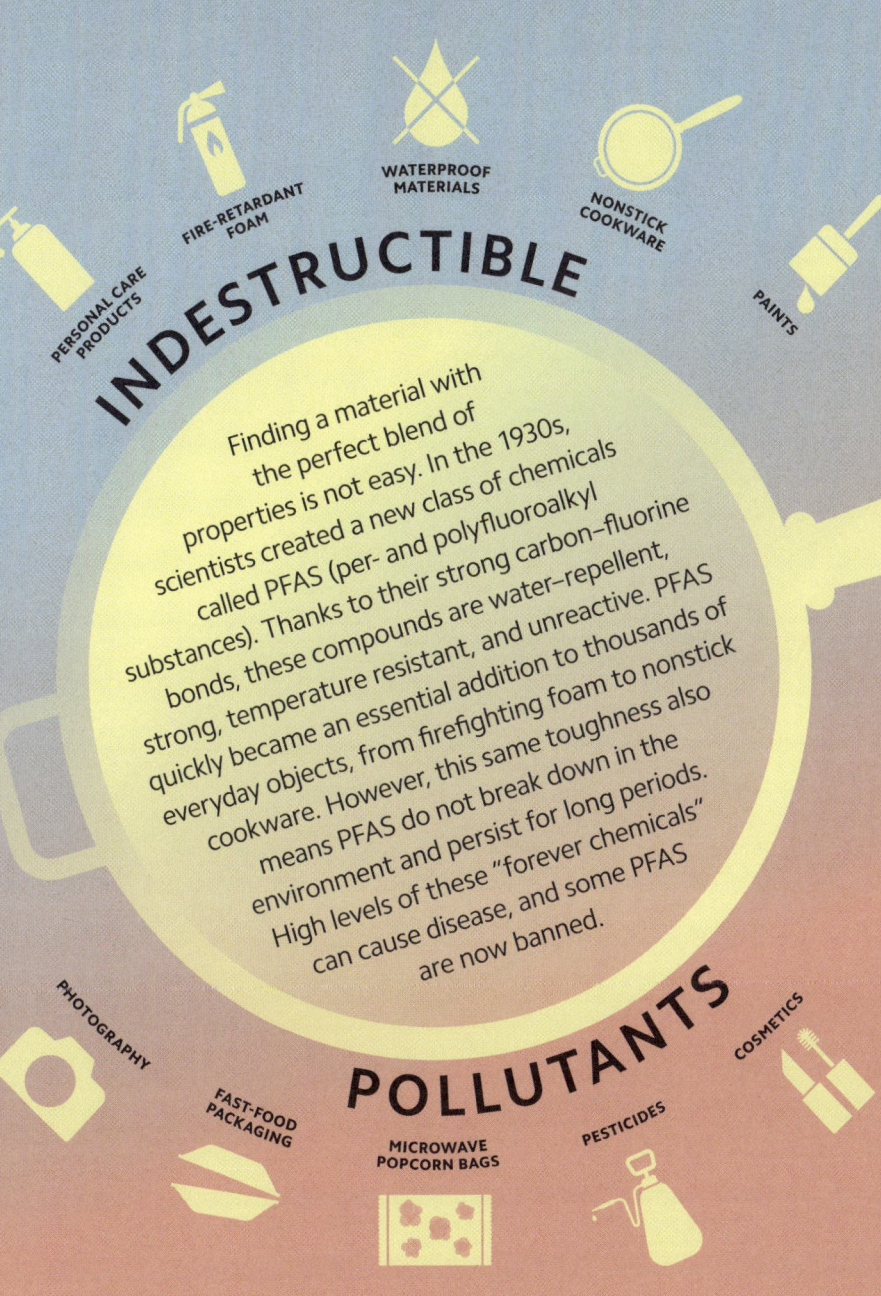

INDESTRUCTIBLE

PERSONAL CARE PRODUCTS
FIRE-RETARDANT FOAM
WATERPROOF MATERIALS
NONSTICK COOKWARE
PAINTS

Finding a material with the perfect blend of properties is not easy. In the 1930s, scientists created a new class of chemicals called PFAS (per- and polyfluoroalkyl substances). Thanks to their strong carbon–fluorine bonds, these compounds are water-repellent, strong, temperature resistant, and unreactive. PFAS quickly became an essential addition to thousands of everyday objects, from firefighting foam to nonstick cookware. However, this same toughness also means PFAS do not break down in the environment and persist for long periods. High levels of these "forever chemicals" can cause disease, and some PFAS are now banned.

POLLUTANTS

PHOTOGRAPHY
FAST-FOOD PACKAGING
MICROWAVE POPCORN BAGS
PESTICIDES
COSMETICS

FOREVER CHEMICALS | 135

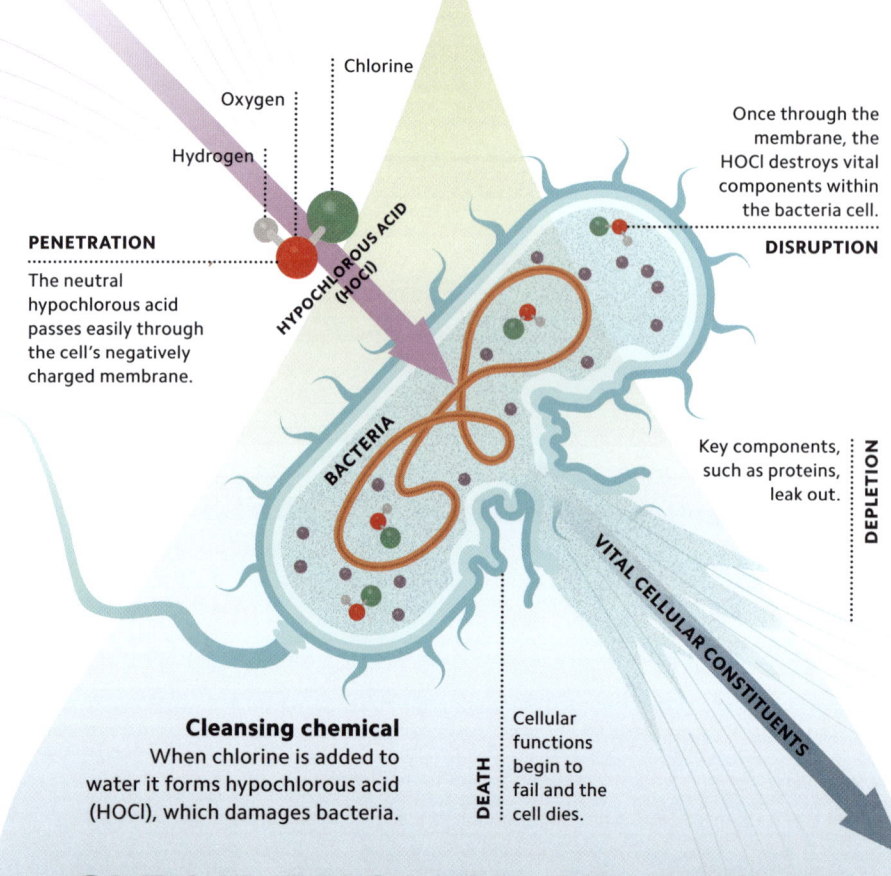

PENETRATION
The neutral hypochlorous acid passes easily through the cell's negatively charged membrane.

Hydrogen
Oxygen
Chlorine

HYPOCHLOROUS ACID (HOCl)

BACTERIA

DISRUPTION
Once through the membrane, the HOCl destroys vital components within the bacteria cell.

DEPLETION
Key components, such as proteins, leak out.

VITAL CELLULAR CONSTITUENTS

Cleansing chemical
When chlorine is added to water it forms hypochlorous acid (HOCl), which damages bacteria.

DEATH
Cellular functions begin to fail and the cell dies.

CLEAN ENOUGH TO DRINK

To make water safe to drink it must first be sterilized. This is a process that involves killing any microorganisms that are present in it. For domestic water supplies, this is achieved by adding chlorine, which poisons any bacteria, viruses, or pathogens present in the water. When added to water, it forms both hypochlorous acid (HOCl) and hypochlorite ions (OCl⁻), both of which damage microorganisms and bacteria. The traces of chlorine left in the water afterwards continue to provide a disinfecting effect.

WATER STERILIZATION

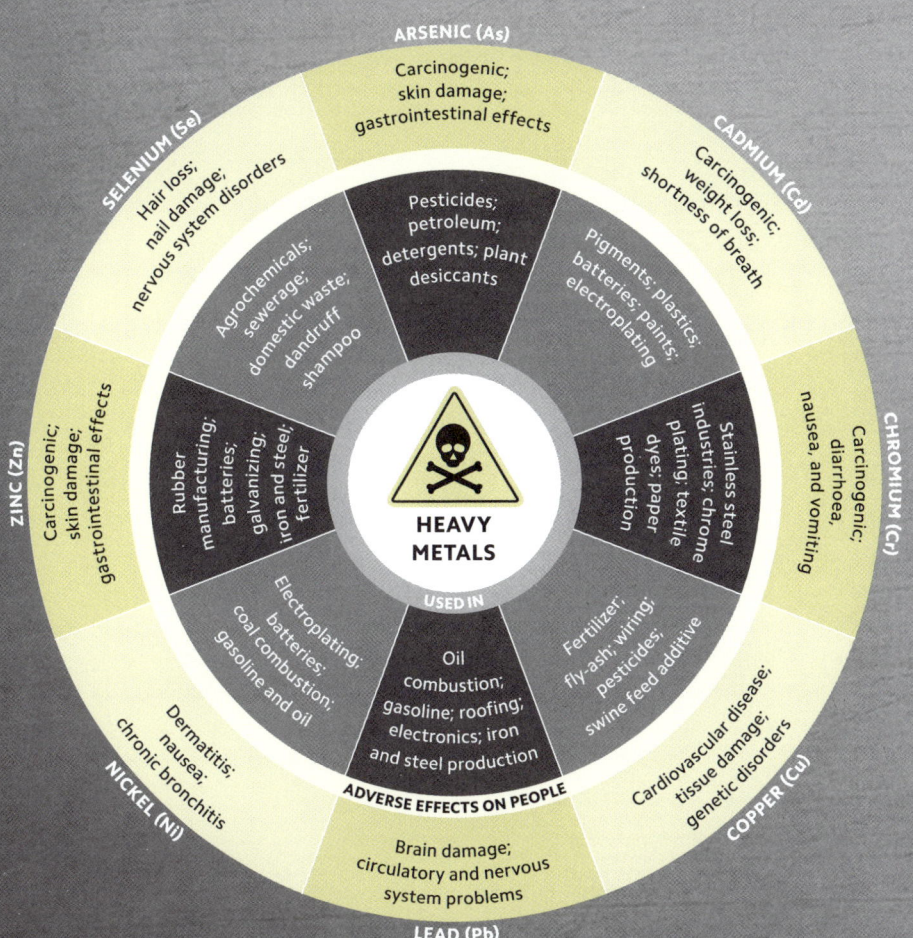

TOXIC BUT ESSENTIAL

Heavy metals are elements from the lower rows of the periodic table (see pp.14–15) with a high atomic weight or high density relative to water. They have many applications in electronics and materials science. However, they can be extremely toxic to living organisms, even at low concentrations, and may accumulate in the body over time. Although naturally occurring, their use is tightly controlled.

NUCLEAR POWER STATION

ATOMIC LEGACY

Splitting atoms releases huge amounts of energy, which nuclear power plants convert to electricity. However, the complex nuclear reactions involved also produce extremely hazardous waste, some of which can remain radioactive for thousands of years. Nuclear plants can extract and use radioactive elements from this waste. But then, radioactive waste material – including spent fuel, contaminated reactor parts, and even clothing – should be stored in sealed containers deep underground. The ongoing radioactive decay (see p.21) generates a lot of heat, so the waste must be cooled in carefully managed conditions during storage.

DEEP UNDERGROUND VAULT

Multiple casks are stored in a single underground vault.

Spent fuel rods are clustered together.

DRY STORAGE CASK

The concrete walls absorb dangerous radiation and provide structural stability.

A fuel rod contains many fuel pellets.

SPENT PELLET

SPENT FUEL ROD

CLUSTER OF FUEL ROD ASSEMBLIES

THICK CONCRETE WALL

METAL LINING

Layers of protection
Hundreds of metres below Earth's surface, nuclear waste is sealed in shielded casks and stored in vaults. This prevents its radioactivity from causing harm to the environment.

138 | NUCLEAR WASTE

TINY PILLS, HUGE IMPACT

More than 90 per cent of medicines are small molecule drugs, which are often taken in tablet form. They are not broken down by the digestive system so can easily reach and diffuse through the small intestine wall to reach the bloodstream. To count as small, the sum total of the atomic weights of all the atoms in a molecule usually comes to less than 500. Such compounds are small enough to pass through cells' outer membranes to target the internal machinery of malfunctioning cells.

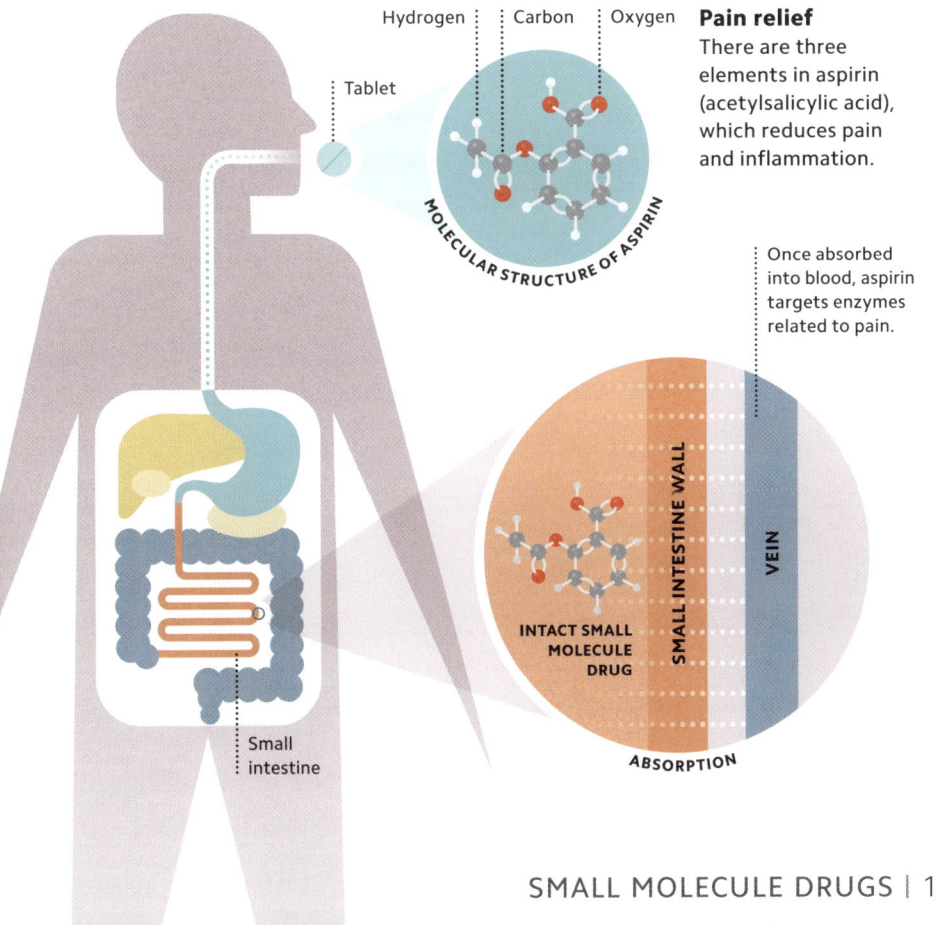

Pain relief
There are three elements in aspirin (acetylsalicylic acid), which reduces pain and inflammation.

Once absorbed into blood, aspirin targets enzymes related to pain.

SMALL MOLECULE DRUGS | 139

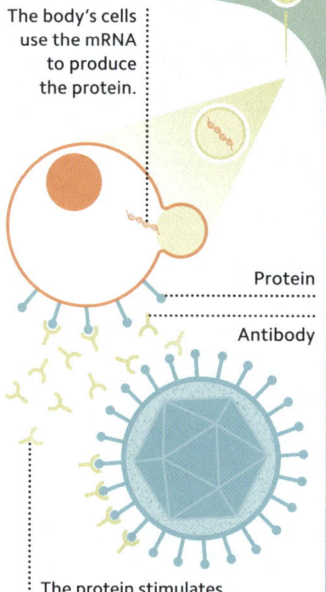

Labs isolate the mRNA sequence for the protein from the virus and then copy it.

A layer of lipids, also called fat, envelopes copies of the mRNA to aid delivery into muscle cells.

VIRUS

mRNA vaccines
Messenger RNA (mRNA) is genetic material that carries instructions for making proteins, such as spike proteins.

The body's cells use the mRNA to produce the protein.

Protein

Antibody

The protein stimulates an immune response – the body produces antibodies that will attack the original virus in case of infection.

DEFENDING WITH CHEMISTRY

When a pathogen enters the body, tiny marker molecules on its surface, called antigens, help the body recognize these infectious particles and trigger the immune system to produce antibodies. If the immune system has never seen a particular antigen before, the person becomes ill while the body learns to respond. Vaccines are chemical instructions for the body to recognize pathogens and create antibodies, without being exposed to the disease-causing parts of the pathogen. If the person later becomes infected, the immune system recognizes the disease antigens and can fight it off. Many vaccines prevent all symptoms.

PRECISION IMMUNE WARFARE

Every cell is covered in tiny individual markers called antigens. These molecules are highly specific to cell types. They help the different types of cell in the body communicate with each other and control processes such as growth and the creation of proteins. It is possible to use these cell signatures to deliver particular treatments, such as cancer drugs, to certain groups of cells. For every antigen there is a matching antibody, and antibody drugs use this specific relationship to carry drugs directly to the targeted cell type, causing minimal side effects.

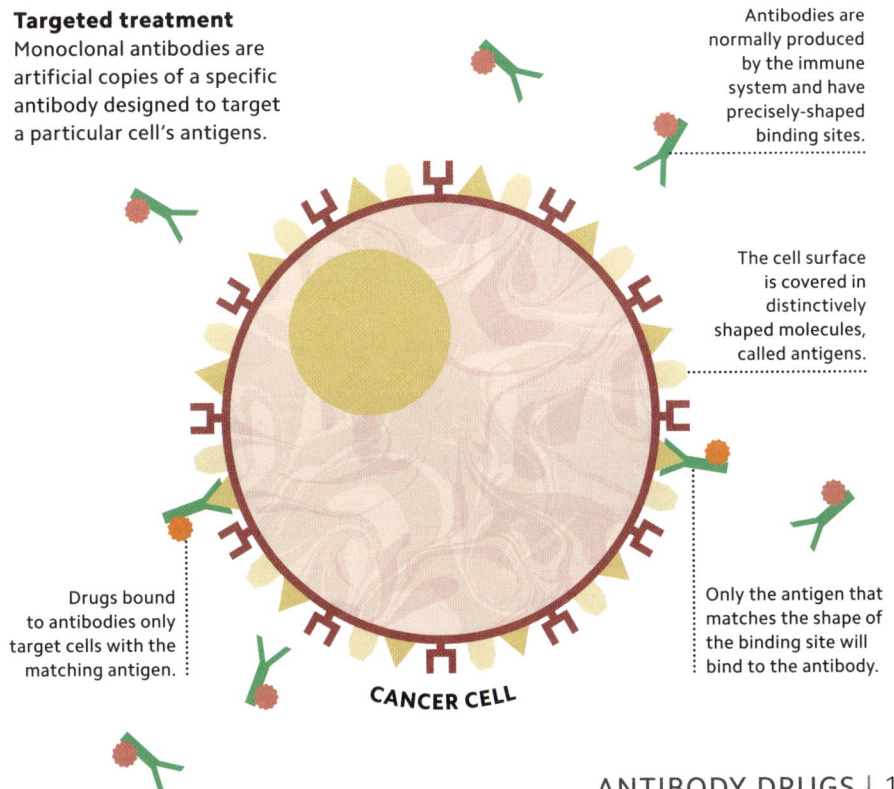

Targeted treatment
Monoclonal antibodies are artificial copies of a specific antibody designed to target a particular cell's antigens.

Antibodies are normally produced by the immune system and have precisely-shaped binding sites.

The cell surface is covered in distinctively shaped molecules, called antigens.

Drugs bound to antibodies only target cells with the matching antigen.

Only the antigen that matches the shape of the binding site will bind to the antibody.

CANCER CELL

ANTIBODY DRUGS | 141

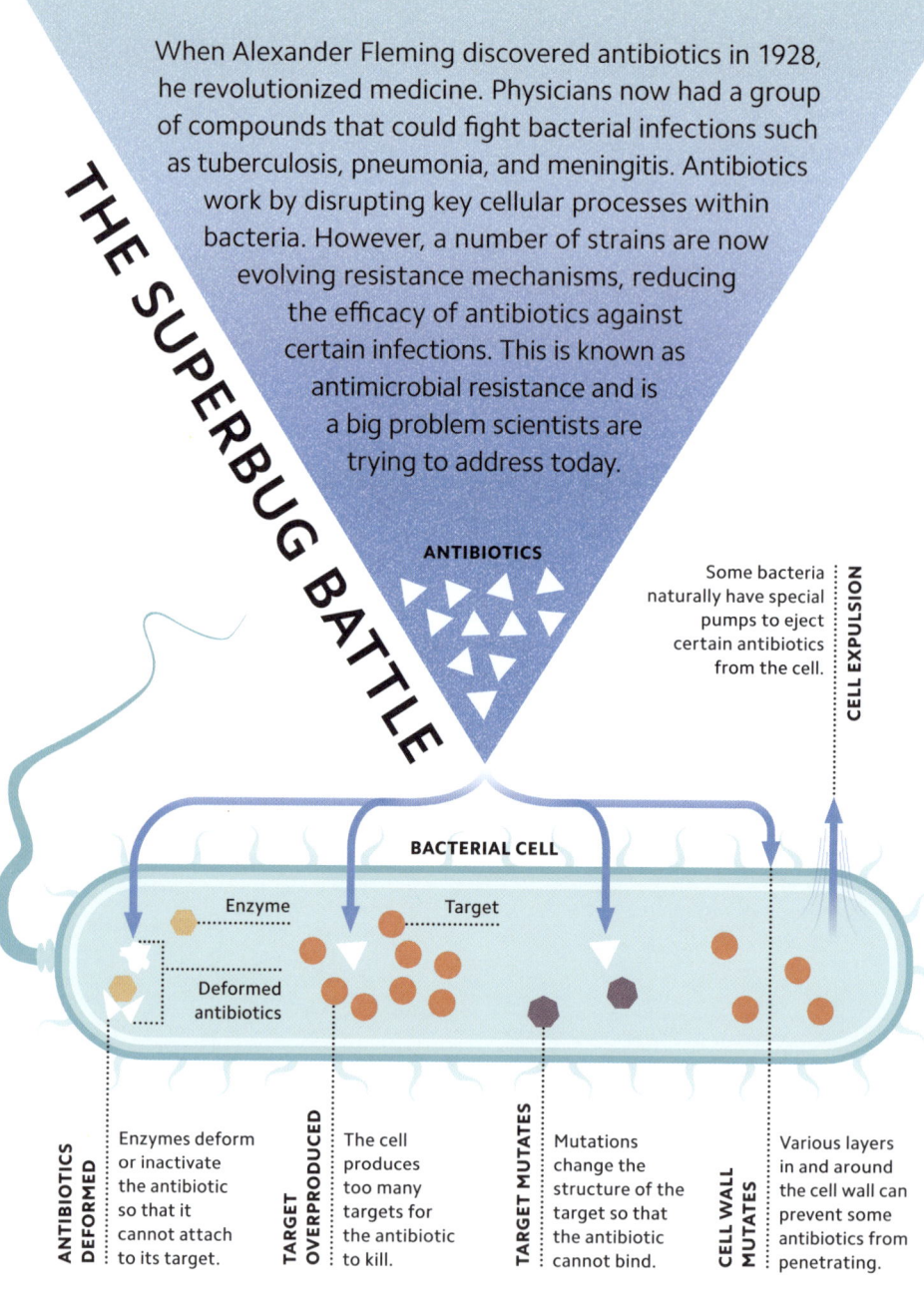

SPLITTING A STRAND

Scientists separate DNA into its two strands. Some of each strand goes into four vessels. They add both normal bases and a small amount of one special base variant (G*, C*, A*, or T*) to each vessel.

COPYING DNA

An enzyme causes a reaction that creates many copies of the test strand, and free bases attach to the copies of the new strands until a labelled base joins.

READING DNA

Strands are sorted by length, and each labelled chain-terminating base is read by a computer.

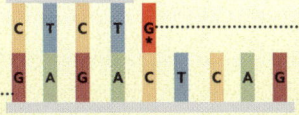

Each vessel contains labelled versions of one base.

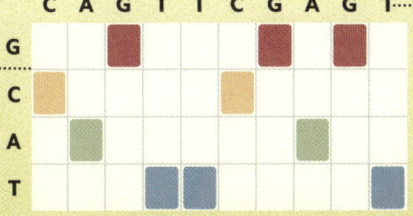

Sanger sequencing

The Sanger method of DNA sequencing reads the order of bases added to DNA as it replicates, one base at a time.

A labelled G stops more bases from joining the strand.

When a labelled base variant joins, the copying sequence ends, creating fragments of DNA of varying lengths.

ENDING A CHAIN

Software combines all the information from the different chains from the four vessels to give the whole sequence.

READING LIFE'S CODE

The order of the four DNA nucleobases – adenine (A), thymine (T), cytosine (C), and guanine (G) – encodes all the genetic information within an organism. Mapping an organism's DNA sequence allows scientists to understand the function of different genes and the effects that different configurations of nucleobases have. There are several different methods of analysis available: some to quickly test short fragments of DNA for specific genes, and more complex tools to sequence entire genomes.

SHOW WITH THE FLOW

Lateral flow tests are a common diagnostic tool, used to quickly detect a particular target substance in a liquid sample. They contain an absorbent pad, coated at one end with molecules called the conjugate that react with the target substance. A chemical reaction with the conjugate "marks" the target substance as it passes. The marked substance sticks to another set of molecules on the test line.

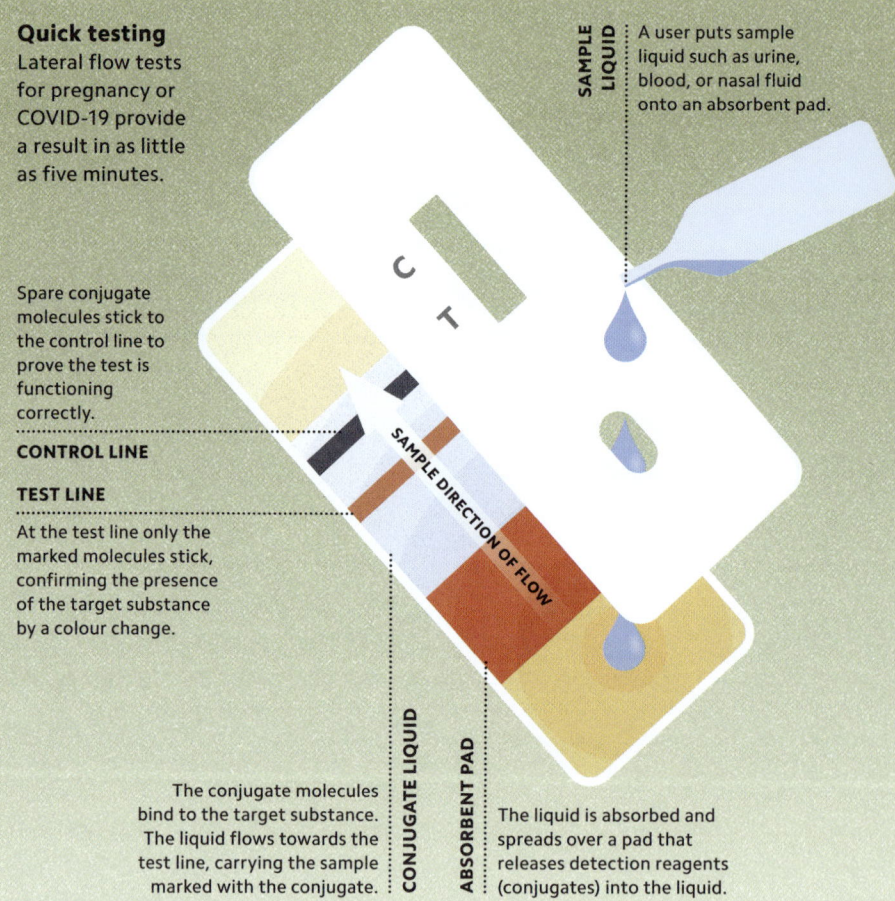

Quick testing
Lateral flow tests for pregnancy or COVID-19 provide a result in as little as five minutes.

SAMPLE LIQUID — A user puts sample liquid such as urine, blood, or nasal fluid onto an absorbent pad.

Spare conjugate molecules stick to the control line to prove the test is functioning correctly.
CONTROL LINE

TEST LINE
At the test line only the marked molecules stick, confirming the presence of the target substance by a colour change.

CONJUGATE LIQUID — The conjugate molecules bind to the target substance. The liquid flows towards the test line, carrying the sample marked with the conjugate.

ABSORBENT PAD — The liquid is absorbed and spreads over a pad that releases detection reagents (conjugates) into the liquid.

Virus vectors
New DNA cannot be inserted directly into a cell. Instead, DNA therapy uses a modified virus to carry a corrected gene into a cell's nucleus.

The new, correct gene is identified and isolated.

The correct gene is inserted into a vector (in this case a modified virus).

The vector is packaged in a vesicle (a soluble, laboratory-made membrane-bound sac) to make delivery easier.

The vesicle penetrates the cell membrane.

In the cell, biochemical machines called ribosomes make a new protein, using the new gene.

The vesicle breaks up and the virus delivers modified DNA to the cell's nucleus.

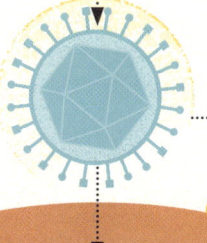

FIXING GENES WITH CHEMISTRY

Mutations or variations in genetic code can interfere with how key molecules in the body are produced, leading to faulty or missing proteins and genetic diseases, such as cystic fibrosis or sickle-cell anaemia. However, DNA therapies can deliver normal copies of specific genes into a patient's cells, helping to restore or modify the function of the affected protein.

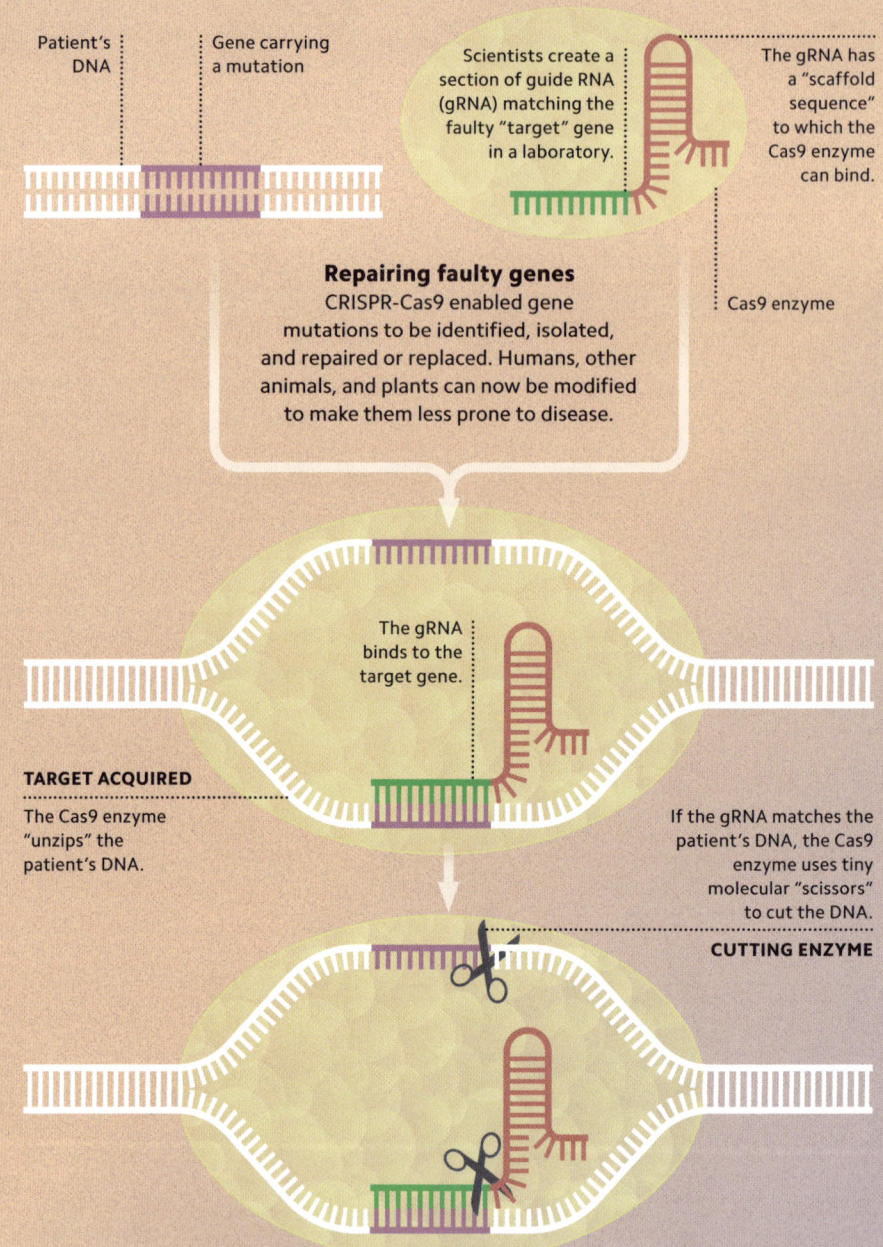

146 | CRISPR GENE EDITING

REWRITING LIFE'S CODE

CRISPR-Cas9 is a gene editing technique developed from the immune systems of bacteria. When a bacterial cell becomes infected by a virus, small repetitive sequences of DNA, known as Clustered Regularly Interspaced Short Palindromic Repeats, or CRISPR, guide an enzyme called Cas9 to attack the viral DNA. Scientists discovered they could harness this combination of CRISPR sequences and Cas enzymes to chop and edit the genes of any organism, essentially rewriting life's code. Researchers are developing CRISPR therapies to cut out genes that cause genetic illnesses and permanently cure patients.

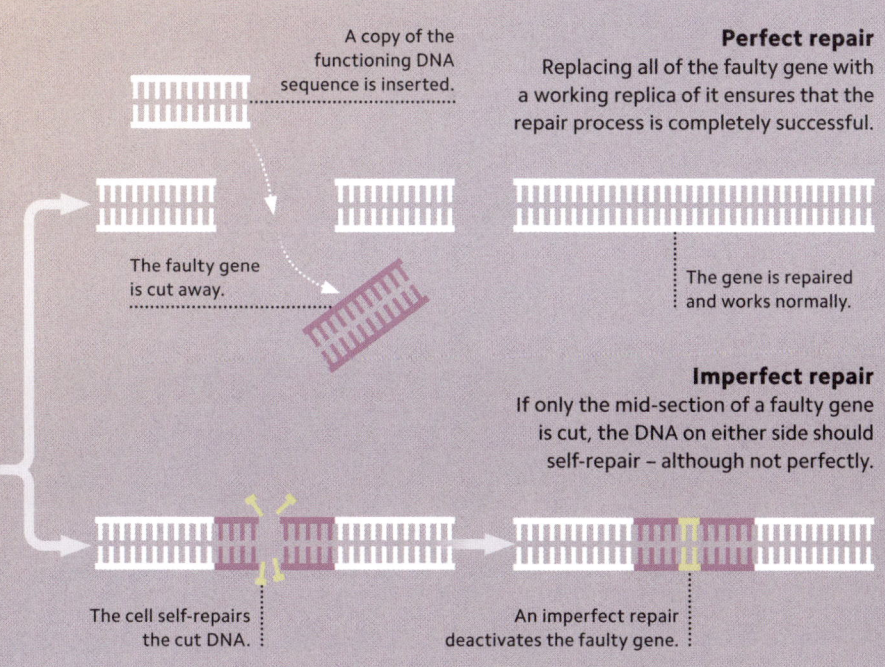

A copy of the functioning DNA sequence is inserted.

The faulty gene is cut away.

Perfect repair
Replacing all of the faulty gene with a working replica of it ensures that the repair process is completely successful.

The gene is repaired and works normally.

Imperfect repair
If only the mid-section of a faulty gene is cut, the DNA on either side should self-repair – although not perfectly.

The cell self-repairs the cut DNA.

An imperfect repair deactivates the faulty gene.

Pure and doped silicon

Pure silicon is a semiconductor. It does not conduct well enough on its own to be useful, so traces of other elements are added to increase its conductivity.

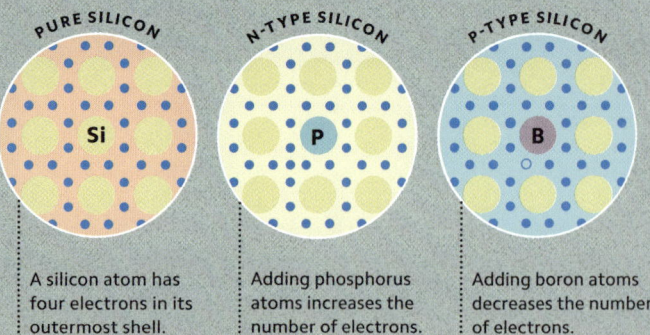

A silicon atom has four electrons in its outermost shell.

Adding phosphorus atoms increases the number of electrons.

Adding boron atoms decreases the number of electrons.

SUPER SWITCHES

Semiconductors are materials that conduct electricity better than insulators (such as glass) but not as well as conductors (such as pure metal). Some substances, such as silicon, are natural semiconductors, which can be "doped" (have impurities added) to fine-tune their electrical properties. Semiconductors are used to make diodes, which control the flow of current, and transistors that act as tiny switches, turning on and off when a small voltage is applied.

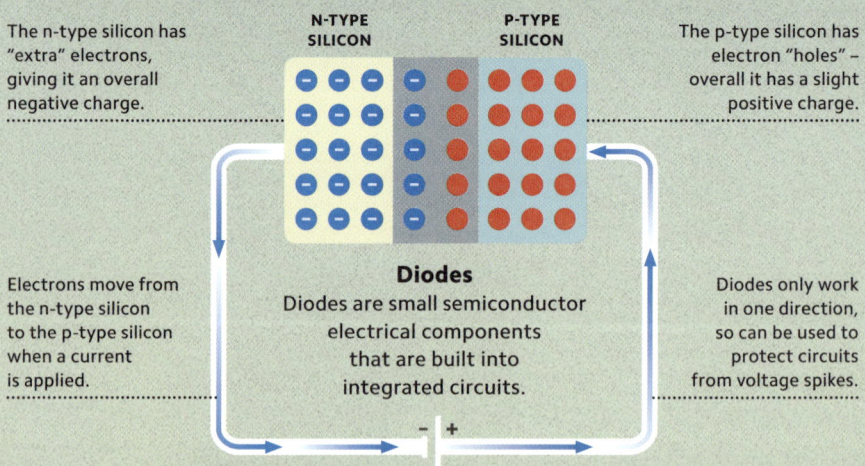

The n-type silicon has "extra" electrons, giving it an overall negative charge.

The p-type silicon has electron "holes" – overall it has a slight positive charge.

Diodes
Diodes are small semiconductor electrical components that are built into integrated circuits.

Electrons move from the n-type silicon to the p-type silicon when a current is applied.

Diodes only work in one direction, so can be used to protect circuits from voltage spikes.

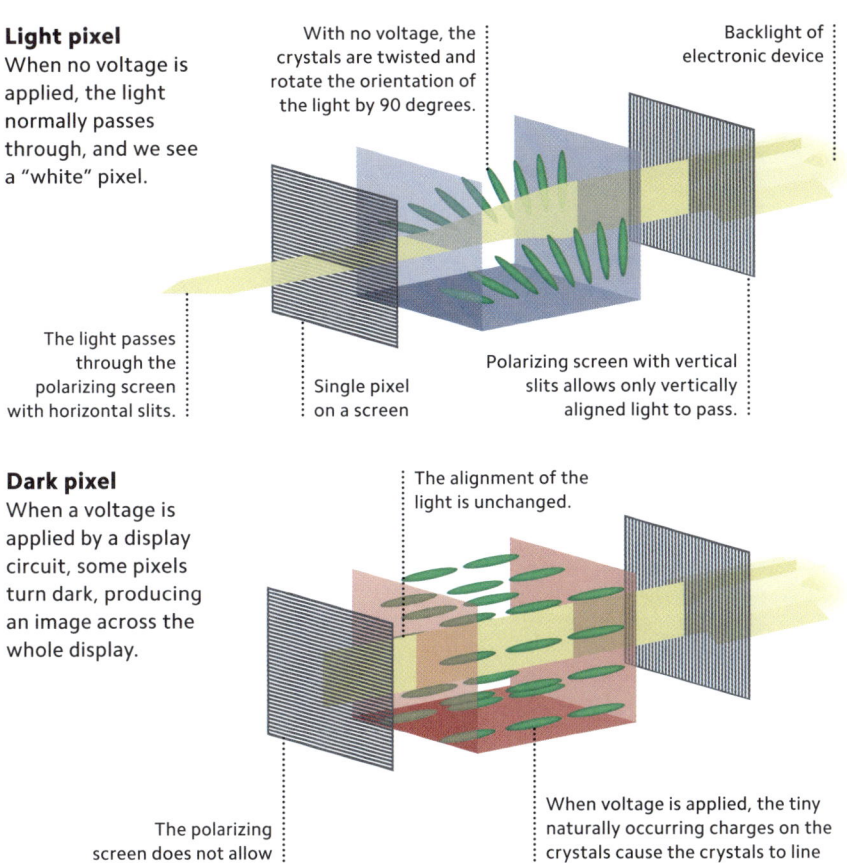

Light pixel
When no voltage is applied, the light normally passes through, and we see a "white" pixel.

With no voltage, the crystals are twisted and rotate the orientation of the light by 90 degrees.

Backlight of electronic device

The light passes through the polarizing screen with horizontal slits.

Single pixel on a screen

Polarizing screen with vertical slits allows only vertically aligned light to pass.

Dark pixel
When a voltage is applied by a display circuit, some pixels turn dark, producing an image across the whole display.

The alignment of the light is unchanged.

The polarizing screen does not allow the light to pass.

When voltage is applied, the tiny naturally occurring charges on the crystals cause the crystals to line up with the electrical field.

DISPLAY STRUCTURES

Liquid crystals are chemical compounds that exist between solid and liquid states – they are usually liquid, but also have properties of crystals. For example, they may flow like a liquid, but their particles are rotated and lined up in an ordered way, like those of a solid. Liquid crystal displays (LCDs) are made up of blocks (pixels), each of which is filled with liquid crystals and can be made transparent or opaque when a tiny electrical voltage is applied.

LIQUID CRYSTALS | 149

BRIGHTER, GREENER SCREENS

SEAL

CATHODE (−)

ELECTRON CONDUCTIVE LAYER

HOLE CONDUCTIVE LAYER

ANODE (+)

GLASS

POLARIZER

ELECTRONS
High-energy electrons move from the cathode to the emissive layer.

The organic material of the hole conductive layer is different from the electron conductive layer, and light must be able to pass through it.

HOLES
The transparent anode adds holes by removing electrons when current passes through.

In some OLEDs, this layer transports electrons and is made up of organic molecules.

Electrons and holes meet to form light in the emissive layer.

EMISSIVE LAYER

Light particle (photon)

Different organic semiconductors are used to produce red, green, and blue light. The chemical properties of each semiconductor determine the colour, brightness, and efficiency of the light produced.

LIGHT EMITTED

Thinner, more flexible screens in mobile phones, laptops, and televisions are emerging because of organic light-emitting diodes (OLEDs). OLEDs are built from layers of organic semiconductors (see p.148). When voltage is applied, high-energy electrons enter on one side, and low-energy electrons exit the other, leaving behind holes. These negative and positive charges travel to meet in the middle layer, where they recombine, emitting red, blue, or green light. Unlike LED and LCD displays (see p.149), OLED screens need no backlight, making them more energy efficient.

150 | ORGANIC LIGHT-EMITTING DIODES

TINY TECH, HUGE IMPACT

Quantum dots are minute crystals of semiconducting materials designed and manufactured using chemical processes to have specific properties. Between 2 and 10 nanometres in diameter, quantum dots are so small that their electrons are confined to a very restricted physical space and adopt discrete energy levels. The gap between these levels can be tuned for particular applications, making quantum dots important in nanoscale science, including applications such as targeted drug delivery.

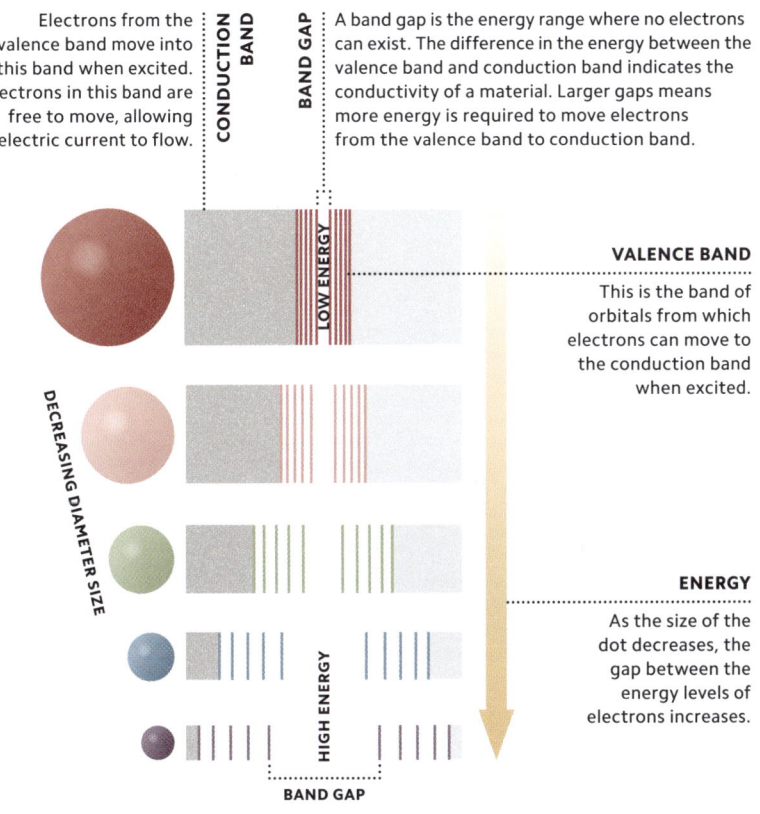

CHEMICAL SCRUBBING

The waste gases released by factories and power plants are fuelling the climate crisis (see pp.120–121). Chemical scrubbing is a process that can help reduce emissions of greenhouse gases. The exhaust gases are passed through a liquid or solid solvent that selectively absorbs or dissolves the pollutants. These pollutants are then stripped away and stored or reused.

Scrubbing carbon dioxide
An amine solution, such as monoethanolamine (MEA), scrubs out the carbon dioxide (CO_2).

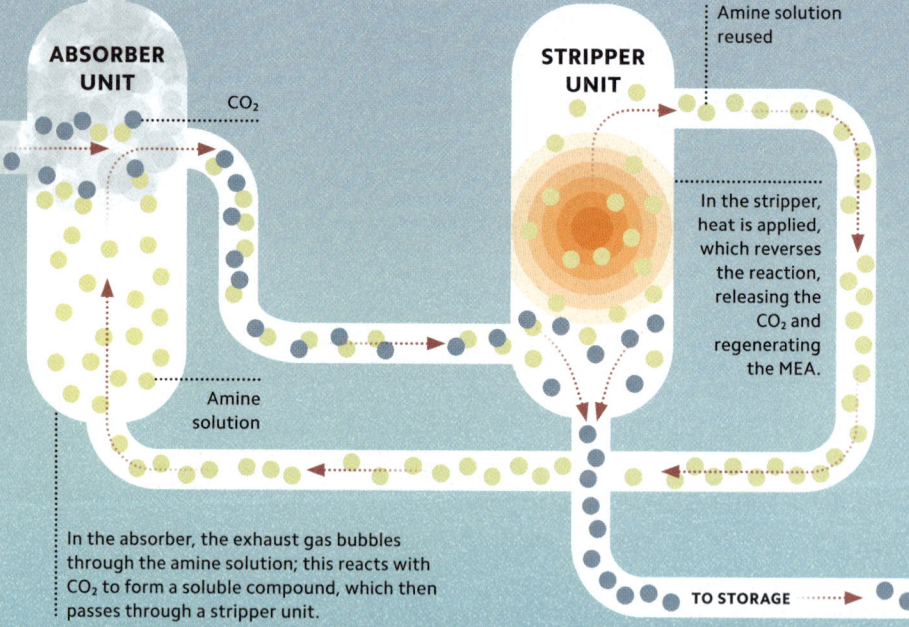

Clean gas, stripped of CO_2, is released into the atmosphere.

ABSORBER UNIT

CO_2

Amine solution

In the absorber, the exhaust gas bubbles through the amine solution; this reacts with CO_2 to form a soluble compound, which then passes through a stripper unit.

STRIPPER UNIT

Amine solution reused

In the stripper, heat is applied, which reverses the reaction, releasing the CO_2 and regenerating the MEA.

TO STORAGE

152 | REMOVING EXHAUST GASES

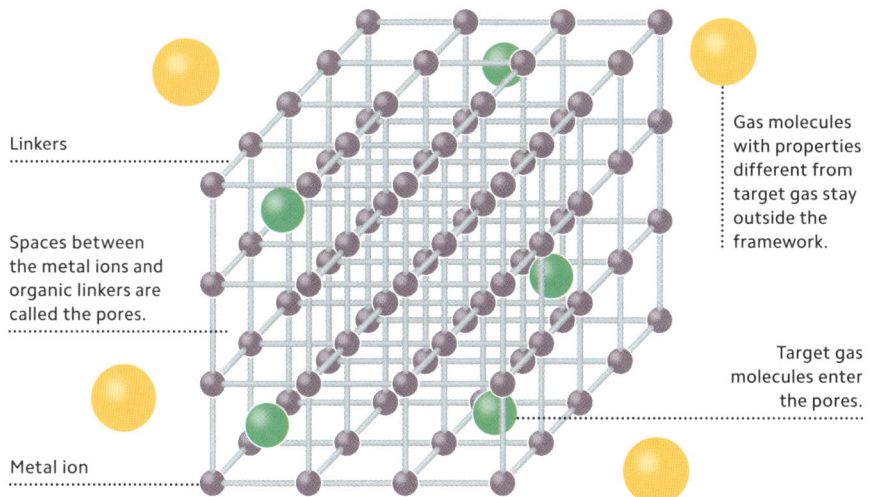

Crystalline materials
MOFs are a lattice of clustered metal ions joined by a variety of organic molecules, also called linkers or struts. The target gas molecules stick to the surfaces inside the MOF's pores through interactions with organic molecules.

TINY HOLES, BIG USES

Crystalline porous materials – substances with billions of tiny holes – have enormous surface areas. They can absorb gases, such as hydrogen or carbon dioxide, into their nanostructure, soaking up molecules like a sponge. Scientists are exploring ways to tailor porous materials and polymers to absorb specific gases. For example, metal-organic frameworks (MOFs) are polymers with adaptable composition and sizes of pores, in which gas molecules can be selectively trapped.

POWERING THE FUTURE

A lithium-ion battery is a kind of rechargeable battery that generates energy through reduction and oxidation, or "redox", reactions in which electrons are transferred from one substance to another. When the battery is in use, lithium atoms in the anode oxidize, releasing electrons as well as lithium ions. The ions travel through the electrolyte to the cathode. When the battery is recharged, lithium ions move from the cathode and embed into the anode again.

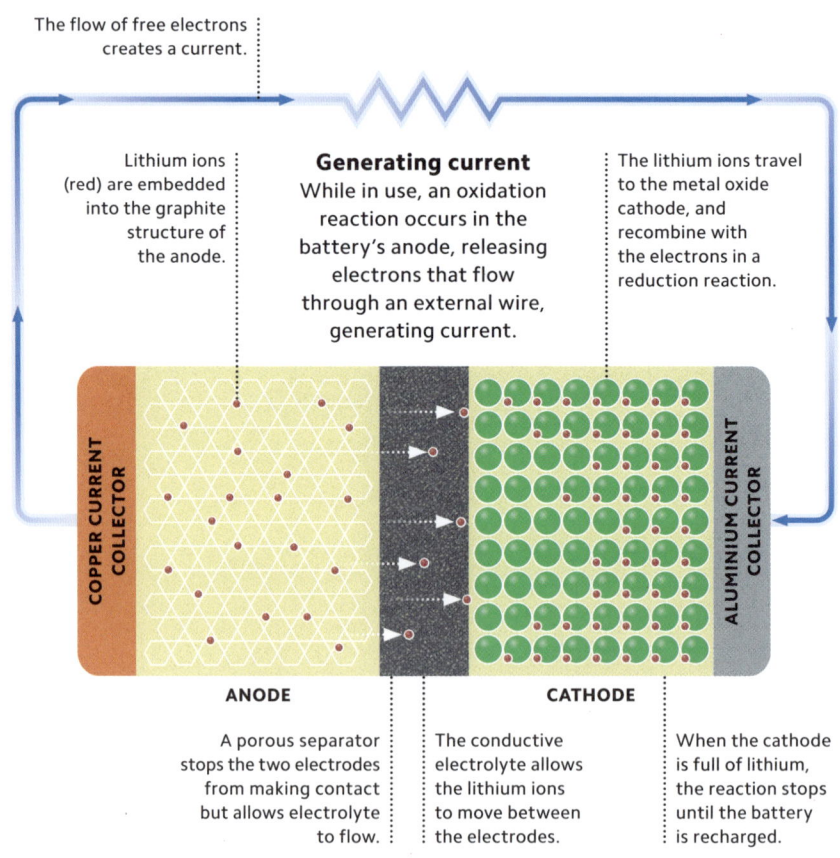

The flow of free electrons creates a current.

Lithium ions (red) are embedded into the graphite structure of the anode.

Generating current
While in use, an oxidation reaction occurs in the battery's anode, releasing electrons that flow through an external wire, generating current.

The lithium ions travel to the metal oxide cathode, and recombine with the electrons in a reduction reaction.

COPPER CURRENT COLLECTOR

ALUMINIUM CURRENT COLLECTOR

ANODE

CATHODE

A porous separator stops the two electrodes from making contact but allows electrolyte to flow.

The conductive electrolyte allows the lithium ions to move between the electrodes.

When the cathode is full of lithium, the reaction stops until the battery is recharged.

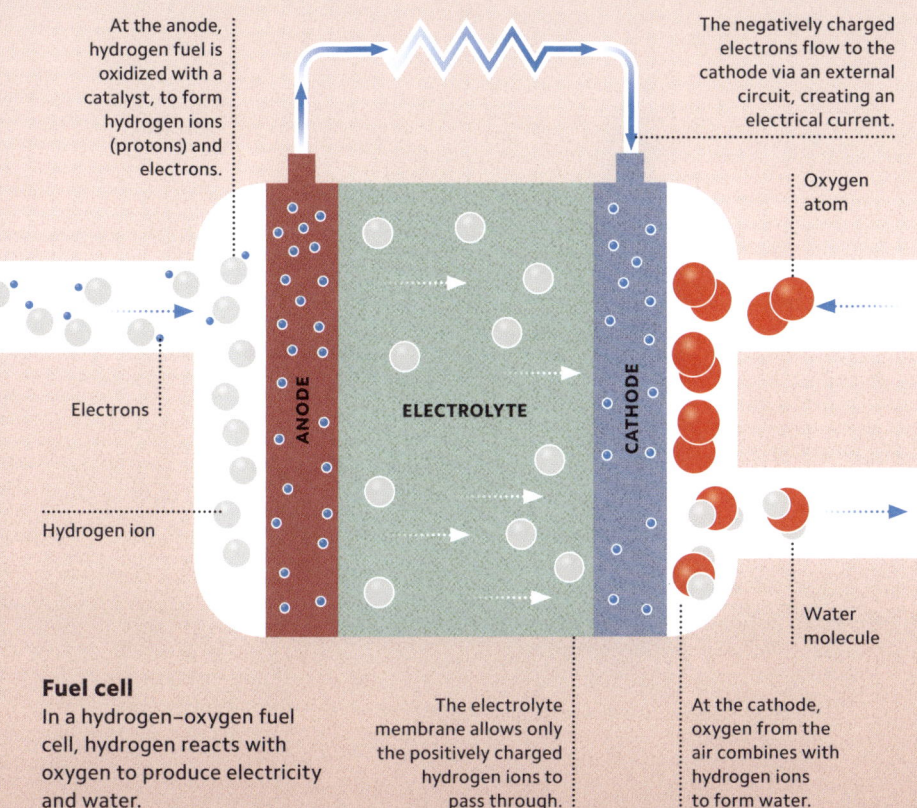

At the anode, hydrogen fuel is oxidized with a catalyst, to form hydrogen ions (protons) and electrons.

The negatively charged electrons flow to the cathode via an external circuit, creating an electrical current.

Oxygen atom

Electrons

Hydrogen ion

Water molecule

Fuel cell
In a hydrogen–oxygen fuel cell, hydrogen reacts with oxygen to produce electricity and water.

The electrolyte membrane allows only the positively charged hydrogen ions to pass through.

At the cathode, oxygen from the air combines with hydrogen ions to form water.

PORTABLE ENERGY

Fuel cells are lightweight devices that produce power electrochemically through two redox reactions between hydrogen fuel and oxygen, releasing water as the only waste product. Fuel cells promise clean electricity but they are low-emission only when they are supplied with hydrogen produced by splitting water through electrolysis (see p.69), using renewable energy. Most hydrogen fuel comes from reacting natural gas with steam, or coal gasification, which releases large quantities of carbon dioxide.

INDEX

Page numbers in **bold** refer to main entries.

A

acids 43, **46**, **47**, 73
actin filament 133
actinides 14–15
activation energy **59**
addition polymerization **72**
addition reaction **64**
adenine 106, 107, 143
ADP (adenosine diphosphate) 107
AI in chemistry **94**
alkali metals 14, **31**, 46
alkaline earth metals 14
alkanes **98–100**
alkenes 75, **98–101**
alkynes **98–99**
allotropes **96–97**
alloys **30**, 67
aluminium 71
amines 73, 102
amino acids 94, **102**
ammonia 56, 57, 62–63, **126**, 127
amorphous plastics **129**
anions 34
anodes 69, 154, 155
antibiotic resistance **142**
antibodies 140, **141**
antigens 141
antimicrobial resistance 142
arenes **48**
argon 35, 119
aromatic molecules (arenes) **48**
aspirin 139
atmosphere **118–19**
atomic force microscopy (AFM) **88**
atoms **8–25**, 28
  allotropes **96–97**
  arenes **48**
  atomic force microscopy **88**
  atomic orbitals **11**, 15
  atomic structure **10**
  chemical reactions 29, 55, 56
  covalent and ionic bonding **36–37**, 95
  cryo-electron microscopy 89
  crystals and gems **45**
  electronegativity **40**
  enthalpy 53
  free radicals 49
  ions **34**
  mass spectrometry 86
  in metals 30
  single, double, and triple bonds **95**
  splitting 138
  stereochemistry **44**
  Van der Waal forces **42**
  X-ray crystallography 90
ATP (adenosine triphosphate) **107**, 113
attraction 10, 42
Avogadro's number 22

B

bases **46**, **47**
batch chemistry 74
batteries, lithium-ion **154**
benzene 48
bioaccumulation 134
biofuels **131**
biosynthetic pathways **112**
blood, haemoglobin **108**
bonds: alkenes and alkynes 99
  atomic force microscopy **88**
  bond enthalpy **53**
  chemical reactions 56
  covalent and ionic bonding **36–37**, 95
  cracking alkanes to alkenes **100**
  vibrational spectroscopy **85**
  X-ray crystallography 90
Briggs-Rauscher reaction 77
bromine 32, 101

C

calcium 20, 24, 37
californium atoms 20
carbohydrates **103**
carbon 14, 15, 16, 17, 68, 96–97
  alkanes, alkenes, and alkynes **98–99**
  hydrocarbons 100, 101, 130
  lipids 104
  in space 123
carbon dioxide 82, 83, 117, 130, 153
  biofuels 131
  catalytic converters 70
  Earth's atmosphere 118, 119
  greenhouse gases and carbon capture **120–21**, 155
  removing 152
carbon monoxide 67, 68, 70
carboxylic acid group 72, 102
catalysts 60, **70**, 72, 74, 105, 126
catalytic converters 70
cathodes 69, 154, 155
cells 107, 141
CFCs (chlorofluorocarbons) **122**
chemical analysis tests **82–83**

156 | INDEX

chemical oscillators **77**
chemical reactions **56**, 58,
   60, 62, 64, 82–83
chemicals: forever chemicals
   135
   in space **123**
chemosynthesis 112
chlorine 29, 32, 37, 83, 136
chlorophyll 117
chloroplasts 117
chromatography **80**, **91**
coal 130
collision theory **60–61**
combustion **68**
compounds 28, 29
concentration 60, 61
condensation
   polymerization **73**
conjugate 144
cooling, endothermic
   reactions 54
copper 30, 66, 81
corrosion **66**
covalent bonding **36–37**,
   95
cracking 99, 100
CRISPR-Cas9 gene editing
   146–47
crude oil 76, 100
cryo-electron microscopy
   (cryo-EM) **89**
crystallography, X-ray 89,
   90
crystals **45**

D

depolymerization 133
diamonds 45, 96–97
diastereomers 44
diodes 148
dipoles 41, 42
disaccharides 103
diseases 142, 144, 145
distillation 38, **76**, 100
DNA (deoxyribonucleic acid)
   94, **106**, 112, **143**,
   145–47
dot and cross diagrams 28
drugs **139**, **141**

E

Earth 17, 25, 122
atmosphere **118–19**, 49,
   120
electricity 69, 148
electrolysis **69**, 71, 132, 155
electronegativity 36, 37, **40**,
   41
electrons 10, 15, 28, 81, 101,
   150, 151
   alkali metals **31**
   arenes **48**
   covalent and ionic bonding
      36–37
   cryo-electron microscopy
      89
   electron capture detectors
      91
   electron clouds **11**
   electron donors and
      acceptors **34**
   electronegativity **40**
   free radicals 49
   halogens 32
   mass of 24
   in metals 18, 30, 34
   in nonmetals 18, 32, 34
   oxidation and corrosion 66
   plasma 13
   semiconductors 148
   single, double, and triple
      bonds **95**
   valence electrons 11, 18,
      30, 31, 33, 34, 36–37, 40
Van der Waal forces **42**
electrophilic molecules 101
elements: alkali metals 14,
   31
   atomic mass 22–23
   balanced equations **29**
   bonding 34
   chemical formulae 28
   compounds 28
   halogens 14, 15, **32**
   makeup of the Universe **17**
   metals and alloys 14–15, **30**
   noble gases **35**
   nucleosynthesis 16
   periodic table **14–15**, 31
   properties of **18**
   radioactive elements **21**
   search for new **20**
   transition metals 14–15, **33**
elimination reaction **64**
emission spectroscopy **84**
enantiomers 44
endothermic reactions 53,
   54, 59
energy 12, 53, 56, 68
   activation energy **59**
   endothermic and
      exothermic reactions **54**
   Gibbs free energy **58**
   enthalpy **53**, 58, 59
   entropy **52**, 58
enzymes 60, 70, **105**, 112,
   134, 142, 147
equations, balanced **29**
equilibria **62–63**
exhaust gases, removing **152**
exothermic reactions **54**, 59
explosions **68**

F

fertilizer 126, **127**
flame tests **81**
Fleming, Alexander 142
flow chemistry **74–75**
fluorine 32, 34, 40
forces, intermolecular 13
fossil fuels 120, **130**, 131
fractional distillation **76**, 100
free radicals **49**
fuel cells 132, **155**
fuels: alternative fuels
   131–32
   fossil fuels 120, **130**, 131
   hydrogen fuel **132**

G

gamma radiation 21
gases **12**, 52, 116
   gas chromatography **91**
   identifying **83**
   porous materials 153
   removing exhaust gases
      152

INDEX | 157

gems **45**
genes 143, 145, **146–47**
Gibbs free energy **58**
glands 109
global warming 120, **130**, 152
glucose 103, 117
glycogen 103
graphite 96–97, 154
greenhouse gases **120–21**

H

Haber process 74, **126**
haemoglobin **108**
half-lives 21, 25
halogens 14, 15, **32**
heat 54, 76
heavy metals **137**
helium 16, 17, 35
herbicides **134**
heterogeneous mixture **38**, 39
homogeneous mixture 38, **39**
hormones **109**
hydrocarbons **98–101**, 104, 130
hydrogen 31, 36, 82, 153
  acids, bases and salts 46
  ammonia **126**
  chemical reactions 56, 57
  DNA 106
  fuel cells **155**
  fuels 131
  green hydrogen **132**
  hydrocarbons **98–99**, 100, 101, 130
  hydrogen bonds **41**
  hydrogenation 74, **75**
  isotopes 19
  lipids 104
  mass of an atom 24
  nucleosynthesis 16
  pH scale 47
  reactivity series 43
  in space 123
  testing for 83
  TNT explosion 68
  Universe makeup 17
  water 28, 41, 55, 116
hydrogenation 98–99
hydrothermal vents 112
hydroxide ions 46, 47

I L

infrared radiation (IR) 85
iodine 32
ionic bonding **36–37**
ions **34**, 86
iron 17, 45, 66, 67, 71
isomers 44
isotopes **19**
lanthanides 14–15
lateral flow tests **144**
Le Chatelier's principle **62–63**
limewater tests **83**
lipids **104**, 140
liquid crystal displays (LCDs) **149**
liquids **12**, 52, 116
lithium 17, 31, 34, 81
  lithium-ion batteries **154**

M N

magnesium 16
mass 22, 24, **55**
mass spectrometry **86**
matter 17
Mendeleev, Dmitri 15
metabolic reactions 105
metal-organic frameworks (MOFs) 153
metalloids 14, 15
metals 14–15, **30**, 37
  alkali metals 14, **31**
  alloys **30**, 67
  electrons 34
  flame tests **81**
  heavy metals **137**
  metal refining **71**
  oxidation and corrosion **66**
  properties of 18
  reactivity series **43**
  reduction from oxides **67**
  transition metals 14–15, **33**

methane 98, 118, 120, 123, 130
microscopy **88–89**
mixtures **38–39**, 80
molecular machines **113**
molecules 44, 88
moles **22–23**
monomers 72, 73, 128, 133
monosaccharides 103
MRI (magnetic resonance imaging) scanners 87
mRNA vaccines 140
muscles 113
myosin head 113
neurons 110
neurotransmitters **110**
neutrons 10, 19, 21
nitrogen 17, 24, 68, 118, 119
  ammonia **126**
  chemical reactions 56, 57
  nitrogen fixation 127
NMR (nuclear magnetic resonance) **87**
noble gases 14, 15, **35**
nonmetals 14–15, 18, **32**, 34, 37
nuclear waste **138**
nucleosynthesis **16**
nucleotides 106
nucleus 10, 31
numbers, scientific notation **24–25**

O

oganesson **20**
OLEDs (organic light-emitting diodes) **150**
orbitals **11**, 15
oscillators, chemical **77**
oxidation **66**, 154
oxides 31, **67**
oxygen 16, 17, 31, 36
  catalytic converters 70
  combustion 68
  Earth's atmosphere 119
  free radicals 49
  fuel cells **155**
  haemoglobin **108**
  lipids 104

oxidation and corrosion **66**
photosynthesis **117**
reduction from oxides **67**
in space 123
testing for 83
water 28, 41, 55, 116
ozone 49, **122**

P

pathogens 140
periodic table **14–15**, 30–33, 35, 40
pesticides **134**
PFAs (perfluoroalkoxy alkanes) 135
pH scale **47**, **82**
phosphate groups 106, 107
phospholipids 104
phosphorus 17
photodiodes 88
photosynthesis **117**
plants 117, 127, 130
plasma **13**
plastics **128–29**, 133
pollutants 135, 152
polyethylene 72, 73, 128
polymerization **72–73**
polymers 72, 73, 128, 133, 153
polypeptides 102
polysaccharides 103
porous materials **153**
potassium 31, 81
precipitates 83
pressure 12, 60, 61, 63
products 29, 56, 64, 73, 74, 105
proteins 94, **102**, 105, 112, 140, 145
protons 10, 15, 19, 21

Q R

quantum dots **151**
quantum theory 11
radiation, gamma 21
radioactive decay 21, 138
radioactive elements **21**
rare earth metals 14–15

reactants 29, 56, 59, 60, 61, 62, 64, 77
flow chemistry 74
limiting reactants **65**
reactions: addition, elimination, and substitution **64**
addition polymerization **72**
chemical oscillators **77**
chemical reactions **56**, 58, 60, 62, 64, 82–83
collision theory **60–61**
combustion and explosions **68**
condensation polymerization **73**
electrolysis **69**
endothermic and exothermic **54**, 59
Haber process 74
and mass balance **55**
reversible reactions **57**, 62
reactive metals 14
reactivity series **43**
recycling **133**
red supergiant stars 16
redox reactions 154, 155
reduction 67, 71
rust 66

S

salts **46**
Sanger sequencing **143**
screens **149–50**
scrubbing, chemical **152**
seawater 38, 39
semiconductors **148–51**
semi-crystalline plastics **129**
semimetals 18
separation 80
silicon 148
sodium 31, 81
sodium chloride 29
solids **12**, 52, 116
solutions **38–39**
solvents 80
space, chemicals in **123**
spectrometry, mass **86**
spectroscopy **84–85**

stars, nucleosynthesis 16
states of matter **12–13**
steel 67
stereochemistry **44**
substitution reaction **64**
substrates 105
surfactants **111**
synaptic cleft 110

T

temperature 12, 58, 60, 63
thermoplastics 128, **129**, 133
thermoset plastics 128, **129**
time, and scientific notations 25
TNT 68
transistors 148
transition metals 14–15, **33**
triglycerides 104

U V W

universal indicator 82
the Universe **17**, 24, 25
UV radiation 49, 122
vaccines **140**
Van der Waal forces **42**
verdigris 66
vibrational spectroscopy **85**
viruses 144, 145, 147
waste 133, 138
water 41, 43, 46, 52
chemical formulae 28
properties of **116**
sterilization **136**
wavelengths 84, 85

X Z

X-ray crystallography 89, **90**
zeolite catalyst 100

INDEX | 159

ACKNOWLEDGMENTS

DK would like to thank the following for their help with this book: Katie John for proofreading; and Vanessa Bird for the index.

Cover images: *Front* and *Back:* DK

DK LONDON
Senior Editor Miezan van Zyl
Senior Art Editor Jessica Tapolcai
Editors Bharti Bedi, Hannah Westlake
Designers Phil Gamble, Clare Joyce
Managing Editor Angeles Gavira Guerrero
Managing Art Editor Michael Duffy
Production Editor Andy Hilliard
Senior Production Controller Meskerem Berhane
Senior Jacket Designer Suhita Dharamjit
Senior DTP Designer Harish Aggatwal
Senior Jackets Coordinator Priyanka Sharma Saddi
Publishing Director Georgina Dee
Art Director Maxine Pedliham
Design Director Phil Ormerod

First published in Great Britain in 2026 by
Dorling Kindersley Limited
20 Vauxhall Bridge Road,
London SW1V 2SA

The authorised representative in the EEA is
Dorling Kindersley Verlag GmbH. Arnulfstr. 124,
80539 Munich, Germany

Copyright © 2026 Dorling Kindersley Limited
A Penguin Random House Company
10 9 8 7 6 5 4 3 2 1
001–355495–Apr/2026

All rights reserved.
No part of this publication may be reproduced, stored in or introduced into a retrieval system, or transmitted, in any form, or by any means (electronic, mechanical, photocopying, recording, or otherwise), without the prior written permission of the copyright owner. DK values and supports copyright. Thank you for respecting intellectual property laws by not reproducing, scanning or distributing any part of this publication by any means without permission. By purchasing an authorised edition, you are supporting writers and artists and enabling DK to continue to publish books that inform and inspire readers. No part of this publication may be used or reproduced in any manner for the purpose of training artificial intelligence technologies or systems. In accordance with Article 4(3) of the DSM Directive 2019/790, DK expressly reserves this work from the text and data mining exception.

A CIP catalogue record for this book
is available from the British Library.
ISBN: 978-0-2417-8433-4

Printed and bound in China

www.dk.com

This book was made with Forest Stewardship Council™ certified paper – one small step in DK's commitment to a sustainable future. Learn more at www.dk.com/uk/information/sustainability